MICHIGAN STATE UNIVERSITY
LIBRARY

AUG 07 2025

WITHDRAWN

PLACE IN RETURN BOX
to remove this checkout from your record.
TO AVOID FINES return on or before date due.

DATE DUE	DATE DUE	DATE DUE
OCT 2 4 1998	FEB 1 1 2002	
	APR 2 3 2002	
DEC 1999	04 30 9 2003	
AUG 1 0 1999	JUN 1 3 2003	
APR 2 1 2004	OCT 1 1 2003	
DEC 0 5 2004		

1/98 c:/CIRC/DateDue.p65-p.14

WETLANDS

Mitigating and Regulating Development Impacts

David Salvesen

 the Urban Land Institute

ABOUT ULI

ULI–the Urban Land Institute is a nonprofit education and research organization that fosters and encourages high standards of land use planning and development. To this end, the Institute sponsors a wide variety of education programs, conducts research, interprets current land use trends, and disseminates pertinent information.

Established in 1936, ULI is recognized as one of America's most respected and widely quoted sources of objective information on urban planning, growth, and development.

Members of the Washington, D.C.–based Institute include land developers, builders, architects, city planners, investors, planning and renewal agencies, financial institutions, and others interested in land use.

David E. Stahl
Executive Vice President

ULI PROJECT STAFF

Frank H. Spink, Jr.	*Staff Vice President, Publications*
J. Thomas Black	*Staff Vice President, Research*
Douglas R. Porter	*Director of Development Policy Research*
David Salvesen	*Author and Project Director*
Nigel Quinney	*Editor*
Betsy VanBuskirk	*Art Director*
Helene Y. Redmond	*Manager, Computer-Assisted Publishing*
Diann Stanley-Austin	*Production Manager*
Jeffrey Urbancic	*Artist*
Laurie Nicholson	*Word Processor*
Laura Smith	*Administrative Assistant*

Recommended bibliographic listing:
Salvesen, David. *Wetlands: Mitigating and Regulating Development Impacts.* Washington, D.C.: ULI–the Urban Land Institute, 1990.

First Printing, March 1990
Second Printing, August 1990

ULI Catalog Number W13
International Standard Book Number 0-87420-697-9
Library of Congress Catalog Card Number 89-52209

Copyright 1990 by ULI–the Urban Land Institute, 1090 Vermont Avenue, N.W., Washington, D.C. 20005-4962.

Printed in the United States of America. All rights reserved. No part of this book may be reproduced in any form or by any means, electronic or mechanical, including photocopying, recording, or by any information storage and retrieval system, without permission of the publisher.

ACKNOWLEDGMENTS

Research for this book involved countless interviews with public officials, environmental consultants, developers, and academics, who provided materials on local, state, and federal wetlands programs or on the case studies that appear in this book. I cannot possibly thank them all, but a number deserve special recognition.

For their assistance in putting together the section on state wetlands programs, I would like to thank Debra Kohne, Florida Department of Environmental Regulation; Charles Newcomb, New Jersey State Planning Commission; Robert Batha, San Francisco Bay Conservation and Development Commission; Les Strnad, California Coastal Commission; Virginia Dodson, Massachusetts Association of Conservation Commissions; Ken Bierly, Oregon Division of State Lands; and Robert Zbiciak, Michigan Department of Natural Resources.

Also, special thanks go to Riley Atkins, King County Department of Parks, Planning and Resources; Marc Boule, Shapiro and Associates, Inc.; and Dyanne Sheldon, (formerly with King County Planning) Jones and Stotes, Inc., for their comments on wetlands development and regulation in King County, Washington, and to John Studt, formerly with the U.S. Army Corps of Engineers, now with U.S. Environmental Protection Agency, for explaining the nuances of the Section 404 program.

Several people deserve special thanks for providing information on various wetlands mitigation projects, including the case studies. In particular Dr. Joe Edmisten, ecologist and private consultant; Ray Pantlik, International Paper Realty Corporation of South Carolina; Jack Wilson, The Wilson Company; Dr. Mary Landin, U.S. Army Corps of Engineers Waterways Experiment Station; Elizabeth Riddle, California Coastal Conservancy; Steve Apfelbaum, Applied Ecological Services; Christos Dovas, Envirodyne Engineers; Wayne Lampa, DuPage County, Illinois Forest Preserve District; Phyllis Faber, wetlands ecologist in San Francisco, California; Philip Williams, Philip Williams and Associates; Thomas O'Laughlin, Morris & Ritchie Associates, Inc.; Douglas Webb, Subdivision Management, Inc.; Ronald Kranz, David Evans and Associates; and Dr. Ron Abrams from Dru Associates.

I would like to give particular thanks to the review committee, William Ishmael, Edward Stone, Mark Viets, Dan Mandelker, and Toby Tourbier, and to Douglas R. Porter, director of development policy research at ULI, for their insightful comments on the manuscript.

Finally, I would like to thank Nigel Quinney, Betsy VanBuskirk, Jeff Urbancic, Helene Redmond, Laura Smith, and Laurie Nicholson, who put the whole thing together.

David Salvesen
Senior Associate
ULI–the Urban Land Institute

TABLE OF CONTENTS

INTRODUCTION	1
CHAPTER 1: THE NATURE OF WETLANDS	9
What is a Wetland?	9
Wetlands Types	9
Wetlands Values	14
Wetlands Losses	18
CHAPTER 2: FEDERAL WETLANDS REGULATION	21
Federal Laws Affecting Wetlands	22
Permitting under Section 404 of the Clean Water Act	28
The Mitigation Dilemma	32
The Takings Issue: Wetlands Protection Versus Property Rights	34
Advance Identification	38
CHAPTER 3: STATE WETLANDS REGULATION	43
State Assumption of the 404 Program	43
Indirect Programs	44
Direct Programs	46
Six Examples of State Wetlands Programs	50
Florida: All in the Coastal Zone	50
New Jersey: Innovative Approach to Wetlands Regulation	52
California: Two Programs for the Coast	55
Oregon: Emphasis on the Estuaries	59
Michigan: Running Its Own 404 Program	62
Massachusetts: Regulating Wetlands under Home Rule	64
CHAPTER 4: MITIGATION STRATEGIES	69
Mitigation Choices	69
Avoidance/Minimization	70
The Villages of Thomas Run	72
Montecito Apartment Village	74
Restoration	77
Bayport Plaza	82
Lakeview Corporate Park	85

	Enhancement	89
	Brookshire Estates	90
	Haig Point	92
	Creation	95
	The North–South Tollway	99
	Westford Corporate Center	103
	Guidelines for Successful Mitigation	105
CONCLUSION		**109**
BIBLIOGRAPHY		**113**
APPENDIX:	**COMMON AND SCIENTIFIC NAMES OF PLANTS MENTIONED IN TEXT**	117

INTRODUCTION

Some of the greatest U.S. cities would not be what they are today if wetlands had not been drained, filled, and developed. Washington, D.C., was built on a swamp, so were substantial portions of New York City, Philadelphia, New Orleans, San Francisco, and Boston. Many of Boston's scenic hills were leveled to fill the wetlands below. Even the renowned Beacon Hill was sacrificed to satisfy the city's appetite for more land. Around Seattle, what could be filled was filled. Land poor but people rich, these cities faced constant pressure to expand; and as their populations swelled, development spilled over into nearby wetlands—a trend that continues today.

Filling wetlands served two important social functions: it created additional land for development, and it rid a city of an apparent nuisance. For centuries, wetlands were considered a source of disease and pestilence, a quagmire of precarious earth and bloodthirsty insects. John Bunyan's *The Pilgrim's Progress*—written in the 17th century and widely read throughout the 18th— describes a bog called the Slough of Despond as an "impending obstacle, a despairingly low point, a clogged encumbrance, on the road to the promised land. . . ."[1] Motivated by similar sentiments, Congress in the mid-1800s granted millions of acres of swamps to the states with the expectation that the new owners would convert what were considered vast wastelands into productive areas, like farmland. Thousands of acres of wetlands in the farmbelt were subsequently drained for agriculture. A few states lost over 90 percent of their wetlands.

> *Let the waters under the heaven be gathered into one place, and let the dry land appear: and it was so.*
> — Genesis 1:9

This century, in critiquing Pierre L'Enfant's plan for Washington, D.C., Lewis Mumford wrote that L'Enfant had "made the most of what was, before the hand of man touched it, a discouraging site: bottom land, bordered by a swamp on the Potomac side, and dissected by a small river. . . ."[2]

Today, however, things have changed. After centuries of mistreatment, wetlands have shed their dismal image and are finally valued for their enormously important environmental and economic functions, such as flood control and the provision of wildlife habitat. Those who traditionally drained and filled wetlands, such as farmers, developers, and engineers, now venture to undo the damage inflicted years ago.

All across the United States, wetlands are making a comeback. Examples abound. On the West Coast, dikes that were built decades ago to convert wetlands to farmland or salt ponds are now being removed. After years of separation from the natural ebb and flow of the sea, the wetlands are slowly returning. Likewise, on the East Coast, restoration is underway in wetlands that were altered and degraded by urban and industrial development. And in the Midwest, a small but growing movement is afoot to convert wetlands that were once drained for farming back to native wetlands and wet prairies, and to stamp out incentives that encouraged them to be drained in the first place.

In Florida, the once-meandering Kissimmee River, channelized by the U.S. Army Corps of Engineers (Corps) into a straight-jacket of a canal, is being given

1

a second chance. In the 1960s, the Corps transformed the lazy, 98-mile river, which ran from Lake Kissimmee to Lake Okeechobee, into an efficient, 48-mile canal. As a result, the river no longer overflowed its banks, and the adjacent marshes, prairies, and swamps dried up and were converted primarily into cattle pastures. Recently, however, after much prodding from the state, the Corps experienced a change of heart. The state convinced it to rejuvenate at least part of the old Kissimmee and its parched wetlands by diverting the water back into the original river channel. The Corps spent over $30 million to take the meanders out of the river, and it will probably cost at least twice that to put them back.

As our understanding and appreciation of wetlands expands, so does the number and scope of federal and state laws to protect them. The laws generally limit certain activities that may destroy wetlands, such as dredging or filling. Since federal laws such as the Coastal Zone Management Act and the Federal Water Pollution Control Act Amendments, commonly known as the Clean Water Act, were enacted in 1972, more and more wetlands have been sheltered under the umbrella of federal protection. For example, under Section 404 of the Clean Water Act, anyone who wants to fill a wetland must first get a permit from the Corps, which usually issues a permit on the condition that the applicant mitigates any adverse impacts on a wetland stemming from the fill. In addition, a growing number of states have adopted wetlands protection laws that are more stringent than federal laws. Ten years ago, only a few states had laws protecting wetlands, now over half the states do, and the list is growing. But the relatively abrupt transformation of wetlands from nuisances to natural treasures has not been smooth, and recent efforts to protect them have met with considerable misunderstanding and resistance.

DEVELOPMENT VERSUS PRESERVATION

Wetlands still face a barrage of threats from real estate and agricultural development. For example, when construction occurs in wetlands, developers usually convert the often mushy wetland substrate into a stabilized building site by filling a wetland with sand, dirt, gravel, or other materials. This not only destroys the filled wetland but may have far-reaching adverse effects on adjacent wetlands as well—for instance, the cutting off of natural circulation and water flow patterns. More commonly, farmers permanently alter the character of a wetland through draining or diking to bring more land into cultivation.

Despite a plethora of laws and programs to protect them, wetlands are likely to encounter increasing development pressures. Demographic trends indicate that people are flocking to water, especially along the coasts where many wetlands lie. In 1984, about 40 percent of the U.S. population lived within 50 miles of the coast. By the mid-1990s, it is estimated that about 75 percent will live near the coast.[3] Along the East Coast in particular, most of the good upland sites have already been developed and many of the remaining sites contain wetlands. In Florida—a state with one of the highest growth rates in the country—most of the remaining undeveloped land contains wetlands. As one Florida consultant observed, "It's hard to walk 200 yards without bumping into a wetland."[4] Around Puget Sound, Washington, another development hot spot, one developer bemoaned that he cannot find a piece of ground that does not contain a wetland. In these and other soggy areas where wetlands predominate, building in wetlands is practically unavoidable, and frequent conflicts between wetlands preservation and development will inevitably occur. Such conflicts arise not just in wet areas, but also in predominantly dry areas that become wet just often enough to be classified as wetlands.

Like a cold Canadian air mass colliding with warm tropical winds, the disparate forces of development and protection have generated a storm of controversy. Why? Because of the incompatible demands placed on wetlands.

In general, landowners and developers prefer to convert wetlands to more economically productive uses, such as houses or shopping centers. They claim that current wetlands protection programs are overly restrictive, unpredictable, and constitute a taking of their property without compensation. Conservationists, in contrast, prefer to preserve wetlands in their natural setting. They too complain about existing wetlands protection laws that, they assert, contain gaping loopholes that allow precious wetlands to be destroyed. By one estimate, the United States loses about 300,000 acres of wetlands every year.[5]

To make matters worse, the relationship between the Corps and the other federal agencies regulating activities in wetlands is often more antagonistic than cooperative. They have yet to implement a clear, consistent policy for protecting wetlands. Furthermore, while state wetlands programs fill some of the regulatory gaps in the federal program, they also inject confusion. For instance, with few exceptions, state wetlands definitions invariably differ, and it is not always exactly clear what constitutes a wetland; a wetland in Rhode Island may not be considered a wetland in Connecticut. Developers will often buy land that is apparently dry only to find out later that, to their chagrin, it is classified as a wetland and is therefore afforded special protection.

Potter's Marsh, one of the most picturesque and productive marshes in Anchorage, Alaska, was created accidentally after the construction of a railroad caused freshwater from mountain streams to collect behind the railroad bed rather than mix directly with the tidal waters of Turnagain Arm. On weekends, throngs of visitors come to observe the many shorebirds that grace the brackish marsh throughout the summer.

THE ROLE OF MITIGATION

Although real estate development accounts for less than 10 percent of overall wetlands losses (agricultural development accounts for most of the rest), the laws against altering wetlands are toughest on developers. Just as developers must pay exactions and impact fees to offset potential impacts of their projects on a local community, or to pay for daycare facilities, low-income housing, or schools, they must also reduce the loss of wetlands and pay for any damages that their projects inflict. Draining or filling wetlands to construct roads, houses, or offices has taken its toll on these natural treasures, but over the past 10 years or so, developers have shown that they can actually improve the quality and increase the quantity of wetlands. By relocating a proposed development to another site to obviate filling a wetland, adjusting lot lines to position most buildings on the upland portion of a site, or creating or restoring a wetland in exchange for any fill, developers have avoided, reduced, and compensated for wetlands alterations. This is generally referred to as wetlands mitigation—a catch-all term for any activity taken to avoid or minimize damage to wetlands, and to restore, enhance, or create wetlands as well.

Wetlands mitigation appears to promise the best of both worlds by allowing development to occur in wetlands while ensuring that wetlands will eventually be "made whole" again. In theory, by avoiding building directly in wetlands, or by creating a new wetland for each one filled for development, mitigation could resolve all conflicts about the development or preservation of wetlands. And although in practice mitigation is often an imperfect compromise, given the strong, competing pressures on wetlands, mitigation is, for now, the best available solution to an otherwise intractable problem. While not a perfect solution nor the technical fix for which many had hoped, wetlands mitigation has significantly reduced the acreage of wetlands destroyed by development and has resulted in thousands of new wetlands being constructed.

Although it shows great promise as a means to stem or even turn the tide of wetlands losses, mitigation has actually fueled the fires of controversy. Arguments arise in part because no strong consensus has materialized on exactly what constitutes mitigation, but also because of disagreements, particularly within the scientific community, on its effectiveness. Many mitigation strategies are still considered experimental; they involve tradeoffs of certain losses for uncertain gains. Wetlands are dynamic ecosystems with unique and often complex geology, hydrology, vegetation, and soils, which make them especially difficult to recreate. Attempts to recreate (or, to use a more fashionable term, "replicate") them have yielded mixed results. For reasons not fully understood, some created wetlands perform marvelously while others do poorly, and it is not uncommon to create an apparent facsimile of a natural wetland only to have it inexplicably fall apart a few years later. Waves can wash away unprotected, fledgling marshes, weeds can invade and quickly overwhelm a site, or desirable wetlands plants may perish because the water is too shallow or too deep. Wooded freshwater wetlands in particular, such as cedar swamps, are extremely complex and have proven stubbornly unpredictable and uncooperative. Some newly created forested freshwater wetlands may take from 50 to 100 years before they begin to resemble a natural system. Who is going to monitor and maintain such long-term projects?

The most controversial form of wetlands mitigation is the creation of a brand new wetland to replace one lost to development. Depending on whom you ask, wetlands creation is either the greatest thing since plastic wrap or it is the biggest hoax since the advent of snake oil. To its supporters, wetlands creation allows development to occur in wetlands, especially where fill is unavoidable, while improving the quality and quantity of wetlands overall. Just as we can grow apples in the desert, these advocates assert, we can create wetlands where none existed before. Critics of wetlands creation, however, declare that there is nothing like the real thing: artificial wetlands can scarcely be considered adequate substitutes for natural ones, and until scientists better understand the dynamics of wetlands, these complex natural systems should not be moved around the landscape like checkers on a checkerboard. Man, the critics argue, cannot successfully create wetlands; only God can. According to one outspoken botanist, "It is morally unethical to obliterate wetlands without knowing what we're doing."[6] The concept of mitigation is suspiciously viewed as an excuse to alter irreplaceable ecosystems. Each side of what has become an impassioned debate can point to wetlands creation successes and failures to support its argument. The debate rages on in negotiations between developers and regulators, in journals, public hearings, and in state and federal legislatures. In the meantime, development marches on and wetlands filling continues.

Despite the uncertainties, a number of mitigation strategies have been implemented, and although few, if any, tried and true strategies exist, some techniques look promising. Developers are learning to avoid unnecessary wetlands alterations, and a new breed of wetlands scientists is emerging to test, refine, and select those techniques that yield the best results. Although still on the low end of the mitigation learning curve, scientists have unraveled many of the secrets of what makes wetlands flourish, discovered the limits of different techniques, identified the critical factors necessary for successful mitigation, and have now reached the point where they can predict, with some degree of certainty, which mitigation techniques will work under different circumstances, and which will not.

But fundamental questions remain unanswered. When a wetland is filled, for example, what kind of wetland will take its place, and where will it be located? Should the replacement wetland be the same kind as the old one, or will any wetlands type suffice? (In the jargon of wetlands mitigation buffs, this is referred to as either "in-kind" or "out-of-kind" mitigation.) Will it be located on-site, nearby, or 20 miles away?

The U.S. Environmental Protection Agency (EPA) and the U.S. Fish and Wildlife Service (FWS) usually prefer in-kind replacement because they believe it is the best way to restore lost wetlands values. Thus, if a developer fills a salt marsh, an equivalent salt marsh should be created as compensation. It sometimes makes sense, however, to create a different kind of wetland as compensation. Because different wetlands attract different species of animals and plants, wetlands can be tailored-made for the particular species an agency wants to attract. Such "designer" wetlands can, for instance, increase habitat for endangered species. A related issue is whether wetlands should be created on or off site. EPA and FWS usually prefer on-site mitigation, but that is often economically and technically impractical, especially for small fills of one acre or less. Off-site mitigation may be the only reasonable option. One drawback of off-site mitigation, however, is that it often cannot replace the type and functions of the wetland being filled. For example, a created wetland may be too far removed from the filled wetland to provide any value to local animals that relied on it for food and shelter. Ideally, replacement wetlands should be located as close to the original site as possible, and some states require that the created wetland be located within the same watershed as the filled wetland.

MITIGATION BANKING

One promising mitigation technique that recently appeared on the horizon is called "mitigation banking." This technique is especially useful for mitigating small wetlands fills, such as for highway crossings, where individual impacts are typically small and localized but cumulatively significant, and also for large projects, such as marinas, where losses are often unavoidable and the opportunities for on-site mitigation may be limited. Mitigation banking generally works something like this: a degraded wetland is purchased and restored (or a wetland is created) by one party, such as a government agency; this restored site becomes the bank; the values of the restored wetland are somehow quantified and

Revegetation of Bracut Marsh for use as a mitigation bank in Eureka, California, has been slow and patchy.

used as "credits" that can later be withdrawn, at a price, to compensate for unavoidable wetlands fills. The price of the credits covers the cost of acquisition, restoration, and operation. For example, suppose construction of a county highway will require filling one acre of a wetland for an exit ramp. There is nowhere else for the ramp to go and no opportunity to compensate for the loss on site. The county may purchase one acre of mitigation credits from the bank (or two acres if 2:1 mitigation is required) as compensation. Similarly, a developer may restore a wetland and receive a certain number of credits for that restoration. These credits are placed in an account and may later be withdrawn to compensate for unavoidable losses associated with the developer's future projects. The bank can be used to compensate for adverse impacts of one or more development projects.

The attraction of a mitigation bank is that rather than creating small, isolated wetlands to compensate for each wetland alteration, one large, contiguous wetland can be created to mitigate many individual alterations. A single, 10-acre wetland is more valuable and easier to create than 20 unconnected, half-acre wetlands. Also, many developers have neither the experience nor the inclination to restore or create a wetland. It may be more efficient, therefore, to let a resource agency select a site, design and implement a mitigation plan, and be responsible for monitoring and maintaining the wetland. An agency-sponsored bank takes the responsibility and the hassle out of developers' hands and gives it to an agency that will be likely to have a greater interest in the long-term success of the wetland. Mitigation banking can also streamline the permit process for certain sites by providing a readily available supply of credits to those who need and qualify for them. And it gives developers a rough idea of how much mitigation will cost.

One of the main drawbacks of mitigation banking is that it may lead to a net loss in wetlands values if the created or restored wetland in the bank is of a lower quality than the wetland that was filled. Also, banks are susceptible to abuse: developers will have little incentive to reduce wetlands losses if they can simply buy credits from a bank instead. According to one critic, mitigation banking proves that "developers are smarter than regulators," since it allows developers conveniently to buy prefabricated wetlands to mitigate for avoidable losses.[7]

FWS has been involved with 13 mitigation banks over the last several years, with varying degrees of success. The banks' holdings range from 11 to over 9,000 acres. Mitigation banks have been established in Oregon, California, North Carolina, Minnesota, North Dakota, Utah, and Louisiana. Banks have been used most commonly for development of highways and ports.[8]

In Astoria, Oregon, a mitigation bank was established by reclaiming a small piece of intertidal wetland that was diked years ago. Over the last 100 years, more than 86 percent of the tidal marshes in the Columbia River estuary were lost, primarily due to diking. The 32-acre Astoria Airport Mitigation Bank was created by breaching the dike and exposing the site once again to tidal inundation. The bank is used to offset unavoidable wetlands losses incurred by development in the estuary.[9] After correcting some initial problems with the restoration, the tidal wetland is now functioning quite well, according to Oregon's Division of State Lands. But development in Astoria has been slow, and only one debit for 10.5 acres has been withdrawn so far.

In Humboldt Bay, California, a mitigation bank was established in 1981 to offset wetlands losses incurred from filling small "pocket marshes" in the industrial waterfront areas in the city of Eureka. Pocket marshes are small wetlands (of two acres or less) that have been isolated from tidal action by past development, have a relatively low value because of their size and isolation, and cannot be easily restored to a fully functional marsh.[10] Like the Astoria bank, the Eureka mitigation bank was created, at a cost of about $30,000 an acre, by breaching a dike and restoring tidal action to a 13-acre, former wetland that had been filled and used as a lumber yard. Unlike the Astoria bank, however, the Eureka mitigation bank has not been very successful. About five acres were graded and planted to create a wetland dominated by salt marsh cordgrass (hereafter referred to simply as "cordgrass"), but revegetation has been slow, and most of the site remains bare. Nonetheless, the largely unvegetated site is highly attractive to shorebirds and ironically now supports the largest population of Humboldt Bay Owl's Clover, an endangered species that inhabits disturbed sites.[11]

According to an official of the California Coastal Conservancy, the bank was a noble idea that has not worked out as expected. The main problem is that the site is loaded with wood chips, which either float to the surface and drift out to the bay or decompose and create anaerobic conditions that ultimately suffocate desirable marsh plants. In addition, off-road vehicles trample young plants and frighten shorebirds away. The Conservancy has taken a number of steps to improve the condition of the marsh, such as installing a fence to exclude off-road vehicles and revegetating some barren areas. In addition, the Conservancy plans to dredge the tidal channels to improve water circulation and replace the gravel and wood-chip laden, hardpacked soil with cleaner, loamier soil.

RECENT RESEARCH ISSUES

Much of what we know about wetlands derives from research and from the hundreds of wetlands mitigation projects undertaken each year. The trick to establishing mangrove swamps in tropical Florida is to start by creating a salt marsh composed primarily of cordgrass, interspersed with mangrove seed pods. The cordgrass, which often matures in one or two growing seasons, protects the young mangrove seedlings until they become old enough to fend for themselves. Over time, the mangroves replace the cordgrass. Scientists have also discovered that some common wetlands plants are more demanding than others. For example, cordgrass is very picky about where it will grow. It lives only in a very narrow range of elevations and is therefore difficult to establish in a newly created marsh.

Every wetlands mitigation project contributes to a collective understanding of how wetlands function and how they can be successfully created. Over time, experience will weed out the less successful mitigation methods and result in a set of proven techniques for creating artificial wetlands that closely resemble natural ones—a kind of artificial wetlands darwinism. Unfortunately, a number of wetlands creation efforts will fail, some within a year or two, others several years later. In order to prevent, as one wetlands biologist remarked, all the "trials from becoming errors," a number of research efforts are underway to evaluate the long-term viability of created wetlands.

One ambitious research project is being conducted by an organization called Wetlands Research, Inc., along the Des Plaines River in metropolitan Chicago. The site, located 35 miles north of Chicago, incorporates 2.8 miles of the river and about 450 acres of disturbed riparian land. The river had been channelized, and most of its adjacent wetlands were diked and drained for farming or excavated for the gravel beneath. Wetlands Research plans to reconfigure 2.8 miles of the streambed and to reconstruct the wetlands that once naturally occurred. It will construct, at an estimated cost of more than $10 million, eight experimental wetlands areas in which it can control water depth, flow, and detention times. Four of the eight experimental wetlands, which range in size from 4 to 11.6 acres, have been completed. These experimental areas serve as living laboratories where different wetlands creation techniques can be tested. Wetlands Research, Inc., hopes eventually to produce a wetlands design and management operations manual. Through its research, it may be able to eliminate many of the present uncertainties that constrain wetlands creation efforts.

Water levels at Wetlands Research's experimental wetlands along the Des Plaines River will be controlled artificially with pumps and control gates.

Tidal wetlands offer vital resting and feeding places to migrating waders and other shorebirds. Here, curlews and marbled godwits feed at Bracut Marsh.

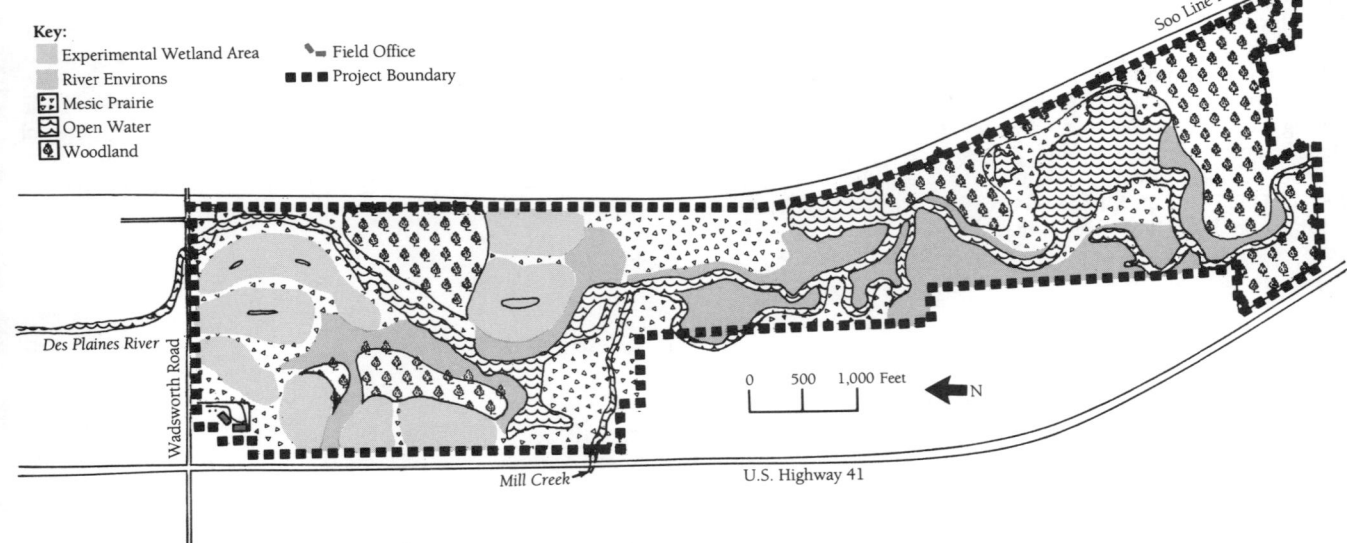

Land cover at the Wetlands Research, Inc., site along the Des Plaines River.

DEVELOPERS AND MITIGATION

Fifteen years ago, few developers were very concerned with wetlands, and the term mitigation had not found its way into their vocabulary. Today, however, one of the first questions developers often ask when buying property is, "Are wetlands present?" Why? Because developers know they will be required to mitigate adverse impacts on wetlands caused by development and that it could easily cost them over $1 million to do so. Developers, however, cannot fairly be expected to shoulder the entire burden of reversing wetlands losses. Although responsible for some wetlands destruction, developers can boast of preserving, restoring, and creating thousands of acres of wetlands. In fact, developers have restored some of the most degraded wetlands simply because no one else is willing and able to pay the sometimes enormous costs. The challenge for regulators is to find ways to improve the overall quality of wetlands while allowing developers the opportunity to obtain a reasonable return on their investment. Developers are, by nature, creative, independently minded people, and where it is in their financial interest to preserve, restore, or create wetlands, they will. Many of their approaches to wetlands mitigation are exemplary, as some of the examples in this book will illustrate.

This book explores many of the current issues in wetlands mitigation, examines federal, state, and local wetlands regulations, describes wetlands mitigation strategies, and provides several examples of recent mitigation efforts. Although only a relatively small, but growing, share of wetlands losses can be attributed to real estate development, this book highlights what developers have done to reduce wetlands losses and to compensate for wetlands fills. Chapter 1 provides a brief background on wetlands: what they are, where they are located, the extent of wetlands losses, and why they are worth saving. Chapter 2 describes federal wetlands regulations, particularly the regulations under the Clean Water Act, and includes an overview of the Corps permit process and a discussion on the takings issue (that is, when, if ever, do wetlands regulations violate the Fifth Amendment of the Constitution by effectively taking private property without compensation). Chapter 3 provides an overview of state wetlands regulations and includes a brief synopsis of six state programs. Chapter 4 presents several wetlands mitigation strategies with case studies provided for illustration. It includes some general guidelines for successful wetlands mitigation based primarily on discussions with regulators and developers.

Notes

1. John Bunyan, *The Pilgrim's Progress*, Part I (1678).
2. Lewis Mumford, *The City in History* (New York, New York: Harcourt, Brace and World, Inc., 1961), p. 405.
3. Testimony of James A. Barnes, U.S. Environmental Protection Agency deputy administrator, given before the Senate Environment and Public Works Subcommittee on Environmental Pollution, 1987.
4. Carol Sawyer Lotspeich, quoted in James Krohne, Jr., "When it Comes to Wetlands, There's Nothing Like the Real Thing," *Planning*, February 1989, p. 5.
5. U.S. Office of Technology Assessment, *Wetlands: Their Use and Regulation* (Washington, D.C.: United States Congress, Office of Technology Assessment, 1984), p. 11

6. Personal conversation on August 10, 1989, with Gerould Wilhelm, research field taxonomist, Morton Arboretum, Lisle, Illinois.

7. Personal conversation on September 25, 1987, with Earl Smith, director of The Wetlands Fund, a private, nonprofit organization raising money for research and development of buffer zones along the Chesapeake Bay in Maryland.

8. See Cathleen Short, "Mitigation Banking," Biological Report 88(41), U.S. Fish and Wildlife Service, Research and Development, Washington, D.C., July 1988, p. iii.

9. See the unpublished report by the Oregon Division of State Lands, "Astoria Airport Mitigation Bank, Astoria, Oregon. Resource Credit Evaluation," November 1986, p. 3.

10. See Short, "Mitigation Banking," p. 49

11. Information from John Zentner, "Wetland Projects of the California State Coastal Conservancy: An Assessment," *Coastal Management*, vol. 16, 1988, pp. 47–67; and from personal conversation on February 15, 1989, with Elizabeth Riddle, program manager, California Coastal Conservancy.

THE NATURE OF WETLANDS

WHAT IS A WETLAND?

In a way, the term "wetland" is a misnomer. Sure, most wetlands have standing or flowing water, but many are dry for part of the year, and others, such as coastal wetlands influenced by the tides, are often dry twice a day. You can often walk in a wetland and not get your feet wet.

No two wetlands are exactly alike: they come in all shapes and sizes, and go by some very unusual names like fens, muskegs, and sloughs. They occur in all 50 states and are as varied as the states in which they occur. They range from infrequently flooded lands to areas that are constantly flooded with deep water. Often found in the transition zone between upland and open water, wetlands tend to be located in depressions or low-lying areas beside lakes, rivers, and streams and along the coast. They are in-between places: not quite land, not quite water. Wetlands usually contain calm, shallow waters and can be as small as a half-acre vernal pool in California, or as large as the broad, shallow, 5,000-square-mile expanse of wetlands that comprise a river of grass called the Florida Everglades—one of the world's biggest marshes. They include isolated prairie potholes that dot the landscape in North Dakota, salt marshes carpeted with thick mats of spartina grass in the coastal plain of New Jersey, extensive bottomland hardwood forests in Mississippi and Louisiana, and mangrove forests that cling tenaciously to the coast of southern Florida. Each has a

> "When I use a word," Humpty Dumpty said in a rather scornful tone, "it means just what I choose it to mean—neither more nor less."
> "The question is," said Alice, "whether you can make words mean so many different things."
>
> Alice in Wonderland

different hydrology, soil condition, and dominant vegetation, and each responds differently to disturbances. Some wetlands are hardy and resilient, while others are more fragile.

Despite their differences, all wetlands have two things in common: they have a soil that is at least periodically saturated or covered with water, and they contain plants that can tolerate such conditions. Water is the controlling factor in determining the type of plant and animal communities that live in wetlands. Most plants cannot survive in the soggy conditions found in wetlands, but others thrive on it. Indeed, the presence of hydrophytic or "water-loving" plants, such as cattail, cordgrass, or pickerelweed, is often used to identify wetlands. Wetlands also exhibit characteristic soil types called hydric soils, which the U.S. Department of Agriculture's Soil Conservation Service defines as soils that become saturated, flooded, or ponded long enough during the growing season to develop anaerobic conditions in their upper layers. The upper soil layers of some wetlands consist of a thick layer of dark, oozy muck, while other wetlands soils develop characteristic patterns of coloration or mottles formed by the presence of different minerals.

WETLANDS TYPES

Wetlands can be divided into two broad categories: tidal and nontidal. This classification, based on water

FIVE EXAMPLES OF WETLANDS DEFINITIONS

U.S. Fish and Wildlife Service

"Lands transitional between terrestrial and aquatic systems where the water table is usually at or near the surface or the land is covered by shallow water. . . . [W]etlands must have one or more of the following three attributes: (1) at least periodically, the land supports predominantly hydrophytes, (2) the substrate is predominantly undrained hydric soil, and (3) the substrate is nonsoil and is saturated with water or covered by shallow water at some time during the growing season of each year."[1]

U.S. Environmental Protection Agency and the U.S. Army Corps of Engineers

"Those areas that are inundated or saturated by surface or groundwater at a frequency and duration sufficient to support, and that under normal circumstances do support, a prevalence of vegetation typically adapted for life in saturated soil conditions. Wetlands generally include swamps, marshes, bogs, and similar areas."[2]

State of Wisconsin

"Those areas where water is at, near, or above the land surface long enough to be capable of supporting aquatic or hydrophytic vegetation, and which have soils indicative of wet conditions."[3]

State of Connecticut

"Wetlands means land, including submerged land, which consist(s) of any of the soil types designated as poorly drained, very poorly drained, alluvial or flood plain by the National Cooperative Soils Survey, as may be amended from time to time, by the Soil Conservation Service of the U.S. Department of Agriculture."[4]

State of California

"Lands within the Coastal Zone which may be covered periodically or permanently with shallow water and include saltwater marshes, freshwater marshes, open or closed brackish marshes, swamps, mudflats, and fens."[5]

Notes:
1. Ralph W. Tiner, Jr., U.S. Fish and Wildlife Service, *Wetlands of the United States: Current Status and Recent Trends*, National Wetlands Inventory Project (Washington, D.C.: U.S. Department of the Interior, Fish and Wildlife Service, 1984), p. 3.
2. EPA, 40 CFR 230.3; and Corps, 33 CFR 328.3.
3. NR 115.03 WAC.
4. 22a–38 Connecticut General Statues.
5. CA Coastal Act of 1976, Section 30121.

---

regimes, parallels many legislative distinctions. Many coastal states, such as California, regulate activities in tidal wetlands only. In addition, few plants grow well in both freshwater wetlands and the salty environment of tidal wetlands. Even cordgrass, which dominates many salt marshes, competes poorly with other plants in freshwater wetlands. Thus, plants provide a convenient distinction between tidal and nontidal wetlands. Tidal wetlands either flood regularly with the daily ebb and flow of tides or only irregularly, during spring or storm tides. Nontidal wetlands, consisting primarily of freshwater wetlands, comprise the bulk (about 95 percent) of all wetlands and occur inland as well as along the coast.

Tidal wetlands include salt and brackish water marshes, mangrove swamps, and intertidal flats. Salt marshes typically occur behind barrier islands and beaches. A handful of soft-stemmed plants such as cordgrass and salt hay along the Atlantic and Gulf Coasts and Pacific cordgrass and glasswort along the Pacific Coast usually dominate such marshes. Cordgrass, a stiff, coarse grass, typically occupies the area closest to the water and is washed twice daily by high tides. Salt hay, a soft, slender grass, grows further from the water's edge. Both plants have developed special mechanisms that allow them to live in harsh conditions of soggy soil and saltwater.

Brackish marshes can be found along the coast where freshwater from rivers and streams meets and mixes with saltwater. They contain wetlands plants such as tide marsh water hemp, pickerelweed, arrow-arum, and cattail among others. Mangrove swamps occur along the

A great white heron peeks above a sawgrass prairie in the Florida Everglades.

coast in tropical Florida and are dominated by either red or black mangroves. Red mangroves occupy the lower, regularly flooded reaches of the swamp and black mangroves live in the upper areas. Unlike most other woody plants, mangroves can live in saltwater. Usually found seaward of tidal marshes and mangrove swamps, as well as at river mouths or along rocky coasts, intertidal flats consist of muddy, sparsely vegetated areas that are regularly inundated by the tides.[1]

Nontidal wetlands generally occur inland along streams, lakes, and ponds and include bogs, swamps, and bottomland hardwood forests. Nontidal wetlands fall into three general categories: emergent, scrub-shrub, and forested.

Freshwater emergent marshes, like their saltwater cousins, are dominated by "emergents"—grasses and grass-like plants that grow partly under and partly above water. Such marshes occur throughout the United States, usually along rivers, streams, lakes, and ponds. The Everglades, a vast, wet area of billowing sawgrass, is a freshwater marsh. Typical freshwater marsh plants include various species of sedges and rushes, such as fox sedge and soft rush, as well as prolific and aggressive plants like cattail, common reed, purple loosestrife, and reed canary grass, which are usually considered weeds. Once these weedy plants invade a site, they can quickly

One of the more aggressive wetlands plants, cattail can form dense monocultures in freshwater wetlands. Although it provides food for wildlife, some biologists rank cattail with other weedy heavyweights such as common reed and purple loosestrife.

SPECIAL ADAPTATIONS OF PLANTS LIVING IN SALT MARSHES

Water is a great equalizer. The principle of osmosis states that when dissolved substances occur in unequal concentrations on opposite sides of a wet membrane, water tends to move from the lower concentration to the higher concentration in order to make the two sides equal. This process spells doom for most plants exposed to seawater, since the concentration of salts in seawater is much higher than that in the cells of most plants, and the saltier water essentially sucks the freshwater, and thus the life, out of the plant.

Relatively few plants can survive the harsh conditions of salt marshes. But some plants, such as cordgrass and mangroves, have developed remarkable adaptations that enable them to thrive in salty, wet environments. For example, cordgrass cells can increase the concentration of salt in their internal water so that it exceeds the concentration of seawater that surrounds them. Freshwater moves into cordgrass cells by osmosis, and any salt that does find its way into the plant is summarily disposed of by glands peculiar to cordgrass and mangroves. These glands secrete a very concentrated solution of salt through special pores leading to the outside of a leaf; water secreted with the salt evaporates and leaves behind salt crystals, which are washed away by the next high tide. Thus, cordgrass survives by keeping most of the salt out of its sap, by secreting what little salt does get in, and by concentrating some of this same salt in its cells so that they will be able to resist the tension placed upon the sap by evaporation through the stomata.[1]

Mangroves, which exhibit some of the same salt management features of cordgrass, have developed a few clever adaptations of their own. For example, red mangrove roots drop down from lower branches of the tree and eventually become stems themselves with their own root systems, while the horizontal roots of black mangroves fan out underground and send up vertical branches through the mud. Both plants create an impenetrable tangle of roots and stems that can crowd out most other plants. Probably the most unusual feature of mangroves, however, is the plant's method of seed dispersal. Red mangrove seeds begin sending out roots even before they fall from the mother plant. After a seed falls, it floats for about a month or so, getting knocked around by the tides, until the root grows sufficiently dense to pull the fledgling plant underwater. With luck, the root will lodge in relatively shallow water and eventually grow into a tangled mass of red mangroves.

Note:
1. John and Mildred Teal, *Life and Death of the Salt Marsh* (New York, New York: Ballantine Books, Inc., 1969), pp. 87–90.

Prairie potholes, such as these in North Dakota, provide vital nesting grounds for over half of all North American ducks.

Freshwater marshes may contain standing water throughout the year, or they may be flooded only a few times during the year. The water level may vary from just a few inches to over two feet deep. Some marshes, such as the prairie potholes in the upper Midwest, may dry out completely during droughts but then bounce back when the rains return. "Potholes"—depressions scoured out of the ground by glaciers during the last ice age—occupy a region roughly 300,000 square miles in size that extends from south-central Canada to the north-central United States. Like the prairie potholes in the Midwest, small, shallow wetlands called "vernal pools" occur in certain regions of central and southwestern California. Little more than shallow depressions that fill with water during the rainy fall and winter seasons and then dry out completely in the summer, these ephemeral, seasonal wetlands support a variety of unusual organisms that have adapted to a life of extremes: bone dry and soaking wet. Water lies just long enough to inhibit the growth of most nonwetlands plants but evaporates before typical pond or marsh plants can flourish. Such conditions favor specialized plants that can either tolerate a wide range of conditions, such as popcorn flower, or that can complete their life cycle during the short span when standing water is present, such as waterwort.[2]

form dense, expansive monocultures and crowd out most other plants. Common reed, which can grow up to 14 feet tall in either fresh or brackish water, is particularly abundant in polluted marshes where little else grows, such as the Hackensack Meadowlands in northern New Jersey.

FIGURE 1.1
CROSS SECTION OF A TYPICAL SALT MARSH

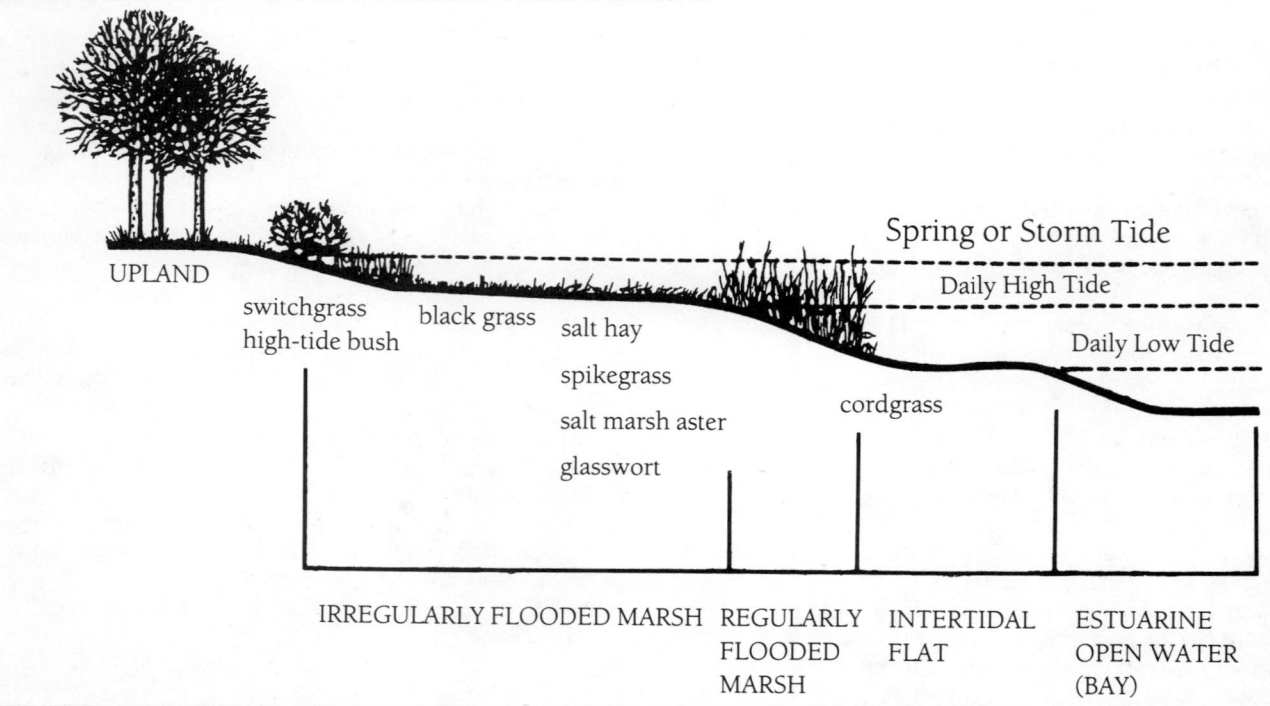

Source: Ralph W. Tiner, Jr., U.S. Fish and Wildlife Service, *Wetlands of the United States: Current Status and Recent Trends*, National Wetlands Inventory Project (Washington, D.C.: U.S. Department of the Interior, Fish and Wildlife Service, March 1984), p. 8.

Scrub-shrub wetlands are dominated by woody vegetation—typically shrubs and small, scrubby trees less than 20-feet tall. They include bogs found in Michigan, Wisconsin, Alaska, and along the East Coast. One of the more unusual wetlands types, bogs are characterized by acidic water and poorly drained, low-fertility soils, often containing a thick layer of partially decomposed plant matter that can reach a depth of over 30 feet. In Pocono, Pennsylvania, such peat soils have long been mined for sphagnum moss. Bogs contain many acid-tolerant plants such as cranberry or bog laurel, carnivorous plants such as pitcher plant and Venus-flytrap. Bogs typically do not experience the regular flooding that typifies marshes and wetlands in floodplains.

A close relative of bogs, pocosins occur primarily in the coastal plain of the Carolinas. Pocosin is an Algonquin Indian word for "swamp on a hill." Like bogs, pocosin wetlands usually contain poor soils, thick layers of peat, evergreen shrubs, and stunted trees. But unlike bogs, pocosins lie in elevated depressions rather than in flat areas along streams.

Forested wetlands include wooded swamps and bottomland hardwood forests that can be found along rivers and streams throughout the United States, although they occur most frequently in the eastern half of the country and in most of Alaska. They contain trees that tolerate prolonged wet conditions, such as willow, red maple, black gum, and northern white cedar in the

Alternately wet and dry, this vernal pool along the Southern California coast hosts an unusual assortment of plants and animals.

north, and bald cypress, tupelo gum, and various oaks in the south. Trees such as bald cypress are noted for their ability to withstand long periods of flooding and

FIGURE 1.2
CROSS SECTION OF FRESHWATER WETLANDS

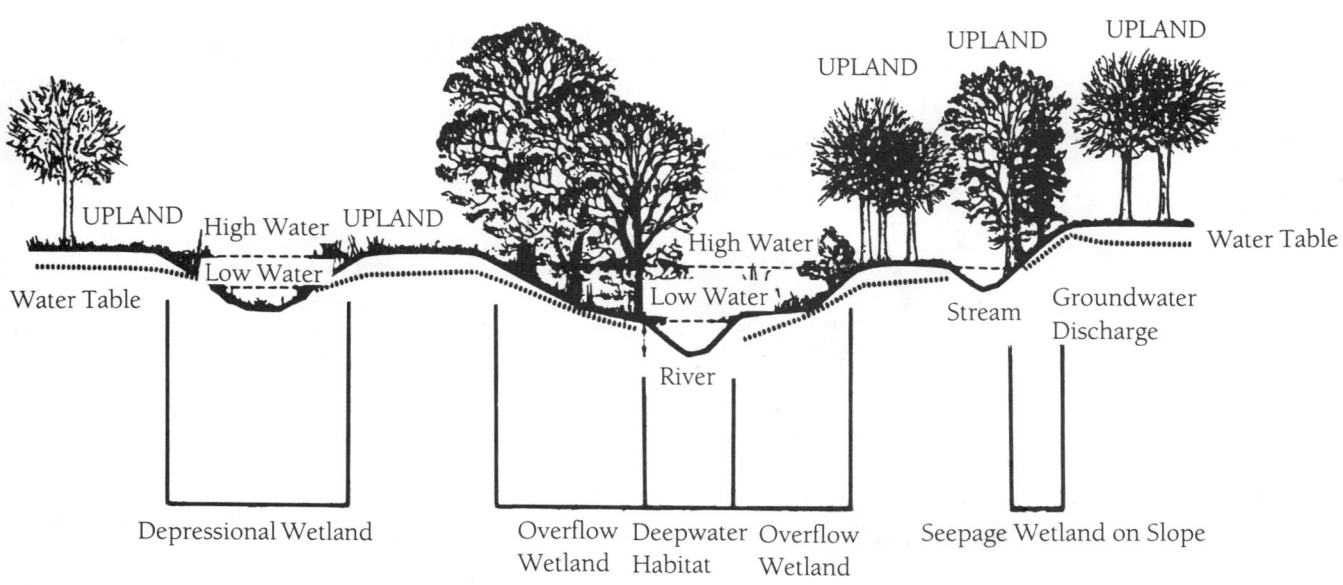

Source: Ralph W. Tiner, Jr., U.S. Fish and Wildlife Service, *Wetlands of the United States: Current Status and Recent Trends*, National Wetlands Inventory Project (Washington, D.C.: U.S. Department of the Interior, Fish and Wildlife Service, March 1984), p. 2.

Bottomland hardwood swamps typically occur in floodplains along rivers and streams. In the Mississippi River valley, such swamps have suffered heavy losses due to conversion to farmland.

gale-force winds. Bald cypress trees, characterized by massive, columnar trunks with swollen, fluted bases, generally contain several descending roots that anchor the tree and many shallow, wide-spreading roots from which rise the peculiar conical structures called "knees." Its root system allows bald cypress to anchor firmly its massive bulk even on the most unstable soils. It is often found in pure stands in bottomland hardwood swamps, such as along the lower Mississippi River valley.

WETLANDS VALUES

Wetlands are what we formerly called swamps, what muskrats dream about, what charlatans in Florida used to sell to unsuspecting retirees in New York. They were considered good only for mosquitoes, ducks, and dumps; a convenient place to put things that nobody else wanted in their neighborhood, like industrial parks, football stadiums, prisons, and airports. Most of the nation's major airports were built on wetlands. Even the names we gave wetlands, like the Great Dismal Swamp along the Virginia–North Carolina border, conjure up images of a gloomy, dreadful wasteland. Wetlands were readily drained, filled, and converted to something more "useful" before we found out that, like fiber, they were good for us. But exactly what are wetlands good for? What unique functions do they serve? Why should they be protected?

Wetlands comprise one of the Earth's most productive natural ecosystems and can out-produce even the most groomed and pampered Iowa cornfields—which is precisely why so many have been drained for agriculture. They have a remarkable knack for capturing and storing sunlight and efficiently recycling materials. They also have an extraordinary ability to shelter fish and wildlife, cleanse polluted and silt-laden water, and protect against floods.

THE FAR SIDE   By Gary Larson

Well, another sucker just bought twenty acres of swampland.

Reprinted by permission of Chronicle Features, San Francisco, CA.

Thousands of acres of coastal marshes along the East Coast were ditched, initially by hand, and drained for agricultural and residential development.

Wetlands provide vital resting, breeding, and feeding places for birds—a pitstop for migrating waterfowl, such as ducks and geese. Over half of all North American ducks nest in the prairie potholes of the north-central United States and southern Canada.[3] In addition to their tremendous natural values, wetlands provide substantial economic benefits as well. Millions of hunters spend a great deal of money—over $300 million per year—to hunt waterfowl.[4] Coastal wetlands, often dubbed the "nursery of the sea," provide essential shelter and spawning grounds for commercially valuable fish like sea trout, blue fish, and flounder, as well as shellfish like oysters, crabs, and shrimp. About two-thirds of U.S. shellfish and commercial sports fisheries rely on coastal marshes for spawning and nursery grounds.[5] In the Southeastern coastal region, for example, over 95 percent of commercial and over 50 percent of recreational fish and shellfish harvests consist of species that depend on estuaries, which are closely linked with coastal wetlands, for all or part of their life cycles.[6] In Louisiana, approximately 75 percent of all commercial marine species, including shrimp and menhaden, rely on coastal marshes and estuaries for part of their life cycle. The state's annual seafood catch is worth about $170 million.

Wetlands help maintain water quality. In some ways, wetlands function like kidneys, filtering out pollutants to purify water before it enters streams, lakes, or oceans. In fact, because of the uncanny ability of wetlands to thrive on nutrient-rich water, a growing number of communities rely on both natural and manmade wetlands to treat tertiary wastewater or stormwater. Houghton Lake, a small community in Wisconsin, pioneered the use of wetlands for wastewater treatment back in 1923. In the Midwest and Southwest alone, over 300 projects receive a total of 10 million gallons of wastewater a day. Most of these wetlands are smaller than 30 acres; the largest, 700 acres, is in Michigan.[7] One of the most innovative and successful examples of using wetlands for wastewater treatment is in Arcata, California. Arcata's mechanical processing plant removes and disinfects solids, which are used as fertilizer in the city parks. The remaining wastewater is then pumped into a 96-acre artificial marsh that filters the

FIGURE 1.3
RELATIVE PRODUCTIVITY OF VARIOUS ECOSYSTEMS

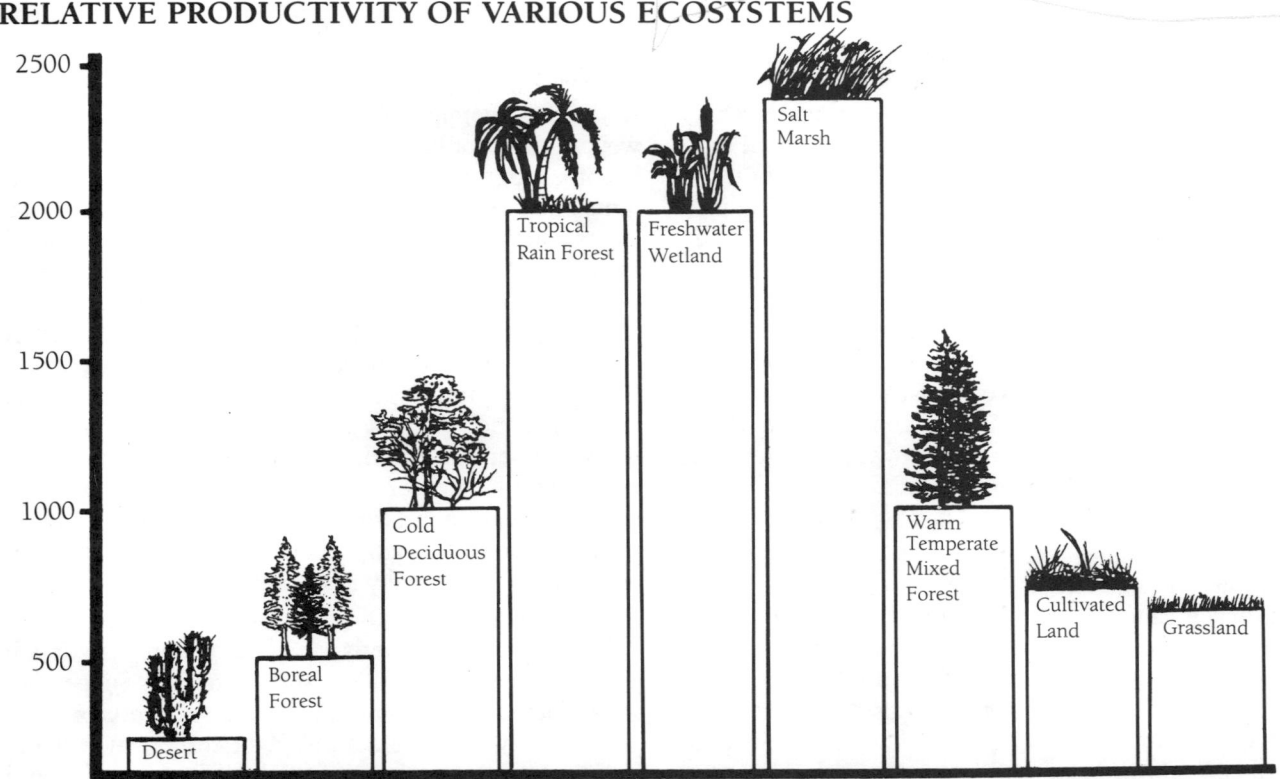

NET PRIMARY PRODUCTIVITY OF SELECTED ECOSYSTEMS [g/m2/YEAR]

Source: Ralph W. Tiner, Jr., U.S. Fish and Wildlife Service, *Wetlands of the United States: Current Status and Recent Trends*, National Wetlands Inventory Project (Washington, D.C.: U.S. Department of the Interior, Fish and Wildlife Service, March 1984), p. 20.

Coastal wetlands protect young fish from predators and provide a rich source of food. Many commercially important fish and shellfish, such as striped bass and shrimp, depend on coastal marshes and estuaries for their survival.

water before it enters Humboldt Bay. The marsh provides not only relatively inexpensive tertiary wastewater treatment but also waterfowl habitat.

Three important properties of wetlands plants make them ideal candidates for treating wastewater. First, their stems and leaves below the water level significantly increase the surface area on which nutrient-eating microbes can attach themselves. Second, wetlands plants transport atmospheric gases, including oxygen, down into their roots, enabling them to survive in anaerobic soils.[8] Third, wetlands convert potential pollutants such as phosphorus and nitrogen into biomass in the form of lush marsh vegetation. Conventional sewage-treatment systems are designed to mimic, in a controlled environment, some of the biological breakdown processes that occur naturally in places like wetlands.

A growing number of developers have come to recognize the advantages of creating wetlands to hold stormwater runoff, either to meet mitigation requirements or as an alternative to typically stark and unattractive retention basins. In Addison, Illinois, for instance, a developer created a 6.5-acre wetland/retention basin to compensate for 5.8 acres of wetlands that were filled to construct an industrial park. The basin contains several zones of vegetation, graded from a deeper zone of open water to the driest area which contains native prairie plants. In between lies an emergent zone of wetlands plants such as tussock sedge and lake sedge (see Figure 1.4).

One of the unique features of this project, designed by Applied Ecological Services, is the biological filtration system used to trap pollutants before they enter the wetland. Generally, stormwater detention facilities and wetlands do not mix because such wetlands eventually become clogged with silt and contaminated with pollutants such as oil, grease, lead, salt, and fertilizers that stormwater runoff carries. But in this case the biological system is designed to catch the silt and pollutants before they enter the basin. The system consists of a sediment trap and, located just downstream, a biological filter "cell" comprised of a shallow wetlands area with dense, emergent vegetation that picks up excess nutrients and gives the runoff water a final polish before it enters the wetland/retention basin. Periodic burning will rid the cell of weeds and some of the contaminants and will actually create conditions favorable to native prairie plants, which have adapted to periodic burning that once occurred naturally in the Midwest.

Despite their ability to breakdown and remove pollutants, however, wetlands can tolerate only so much pollution before they show signs of strain. Suspended sediments can settle and accumulate in wetlands and permanently alter the natural plant composition. High sedimentation rates can literally bury native plants and create conditions favorable to hardier plants such as common reed, which thrives in disturbed wetlands but which supports only a limited level and diversity of wildlife. In 1987, King County, Washington, surveyed

FIGURE 1.4
CREATED WETLAND/RETENTION BASIN IN ADDISON, ILLINOIS

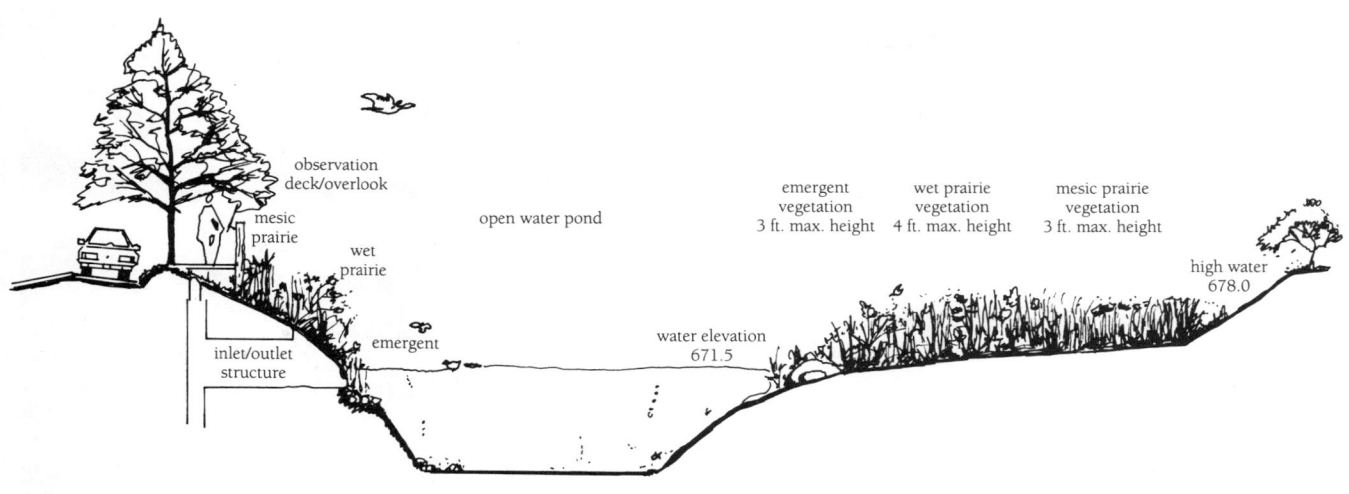

Source: Applied Ecological Services, Juda, Wisconsin.

46 wetlands that had been receiving stormwater runoff from urban areas for an extended period of time. These wetlands were compared to 27 control wetlands—that is, wetlands in undeveloped areas that were not contaminated with stormwater runoff. One of the findings of the survey was that, due to the high levels of sediment carried by stormwater and deposited in most of the 46 wetlands, reed canary grass flourished and dominated many of the affected sites, while the control sites retained their vegetative diversity. As a follow-up to this study, in 1988 King County began a five-year study of the long-term effects of stormwater runoff on wetlands. The study is funded by federal, state, and local agencies.

Wetlands control floods. Like big sponges, wetlands slow down and absorb excess water during storms, then slowly release the stored water and thus reduce peak flows downstream. By slowing the water flow, wetlands allow suspended solids, churned up and carried by floodwaters, to settle and become "subject to the whims and biological urges of bacterial fixation, decomposition, oxidation, volatilization, or uptake by plant roots."[9] Otherwise, the suspended sediments would be carried downstream where they can choke reservoirs, clog rivers, and shorten the life of flood-control basins and dams.

Wetlands do such a good job of controlling floods that they are sometimes preferred over dams. An oft-cited example is the Charles River Natural Valley Storage Project near Boston, Massachusetts, where the Corps decided that the natural wetlands along the Charles River could do a much better job of providing flood control than could a dam. The Corps realized that as the Boston metropolitan area developed and the amount of impervious surfaces increased, the amount of stormwater runoff would increase accordingly. A dam would provide only temporary, and costly, flood control, whereas the roughly 8,500 acres of wetlands would provide effective, inexpensive, natural flood control. The Charles River Project, authorized by Congress in 1974 and completed in 1984, also provides much-needed open space and wildlife habitat next to a major metropolitan area. In the Passaic River Valley in New Jersey, where wetlands that once provided natural flood control have been destroyed by development, flood control will cost an estimated $1.5 billion.[10]

Wetlands can also protect coastal and upland areas from erosion by absorbing and dissipating the impact of waves. Along erosion-prone shores, some communities

On Assateague Island, Virginia, snow geese take a breather from their long migration south.

have established coastal wetlands to control shoreline erosion. Along the Chesapeake Bay, for example, hundreds of landowners now rely on manmade wetlands to prevent their shorelines from being undercut.

Finally, wetlands provide aesthetic and recreational amenities: fishing, hiking, swimming, hunting, and birdwatching. Millions of birds nest in wetlands each summer, or just pass through during their spring and fall migrations; when they do, someone is usually there to watch them. An estimated 50 million people are drawn to wetlands each year to observe and photograph birds.[11]

Only recently have we begun to realize the true value and functions of wetlands and the compelling reasons why they should not be filled indiscriminately. But despite efforts to protect them, many wetlands fall prey to needless disturbances that threaten to destroy their invaluable attributes forever. The problem is often one of scale and timing. Some wetlands make an easy target for development. Filling small, isolated, and seemingly insignificant wetlands may cause only minor, short-term changes in the nearby environment. Eventually, however, all those small wetlands losses begin to add up, causing widespread environmental damage: extensive loss of waterfowl habitat, reduced fish spawning grounds, and more frequent flooding.

WETLANDS LOSSES

The U.S. Fish and Wildlife Service (FWS) estimates that over 50 percent of U.S. wetlands have been destroyed during the last two centuries.[12] According to

Development encroaches on wetlands in California.

FWS, approximately 11 million acres of wetlands—an area over twice the size of New Jersey—in the 48 conterminous states were converted to other uses, such as agriculture, mining, and development, between the mid-1950s and the mid-1970s—an average annual loss of about 550,000 acres. The vast majority of the losses were due to draining and clearing of inland wetlands for farming; urban development accounted for only about 8 percent of the losses.[13]

In coastal wetlands, however, urban development accounts for a much greater share of wetlands losses, particularly in New Jersey, Texas, New York, California, and Florida. In these states, urbanization is responsible for over 90 percent of coastal wetlands losses.[14] Urban development has caused almost two-thirds of Delaware's total wetlands losses.[15] California has lost over 90 percent of its original wetlands. At one time, there were an estimated 200,000 acres of coastal marshes in the San Francisco Bay Area. Today, less than 40,000 acres remain. Many were diked for agriculture or for salt ponds, while others were filled for industrial and commercial development. And the Central Valley wetlands, still the most important winter habitat for waterfowl in North America, have been reduced in acreage by an appalling 95 percent. Before World War II, an estimated 40 to 50 million ducks still wintered in California every year; now the number has dropped to 3 million or fewer. And

White pelicans crowd the beach in Bolivar Flats, Texas.

FIGURE 1.5
WETLANDS CONVERSIONS
(MID-1950s TO MID-1970s)

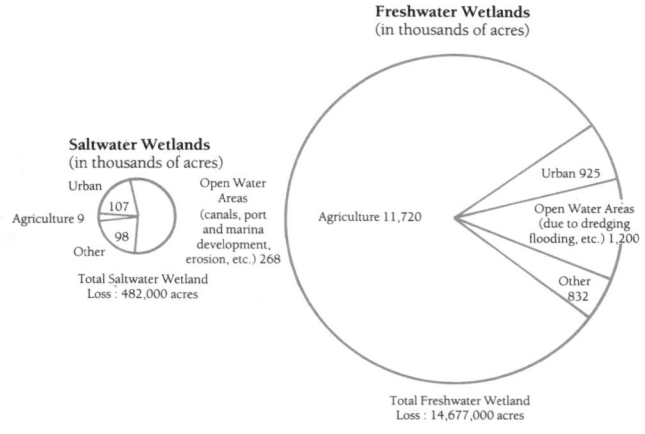

Source: U.S. Office of Technology Assessment, *Wetlands: Their Use and Regulation* (Washington, D.C.: United States Congress, Office of Technology Assessment, 1984), p. 8.

the losses continue—primarily due to dams, diversions, and subsidized irrigation water.[16]

Louisiana loses an estimated 25,000 acres of its coastal marshes each year, mostly through natural subsidence. Historically, Louisiana's wetlands losses were offset by gains, as thousands of acres of wetlands were created each year along the Mississippi River delta. But today, due to both natural and manmade influences, the rate of wetlands losses far exceeds the rate of wetlands gains. Nationwide, the enormous wetlands losses were offset slightly by gains of about 2 million acres from pond and reservoir construction, beaver activity, and from irrigation and wetlands creation projects. On rare occasions, some beautiful wetlands were created by accident. In Anchorage, Alaska, an attractive brackish marsh, which is now part of the Anchorage Coastal Wildlife Refuge, was created in 1917 when the newly constructed Alaska Railroad impeded the normal flow of small mountain streams to the coast. The resultant 600-acre wetland, called Potter's Marsh, which was previously a tidal marsh with adjacent upland habitat, has become Anchorage's most popular place for both shorebirds and birdwatchers.

Prior to enactment of the Food Security Act of 1985 (Pub. L. 99–198, 99 Stat. 1504), federal agricultural policies encouraged farmers to convert wetlands to farmland by providing credit, loans, and commodity price supports. Farmers were encouraged to plant fencerow to fence-row, and wetlands that stood in the way, especially the prairie potholes in the Midwest and bottomland hardwood forests in the Gulf Coast states, were enthusiastically drained, plowed, and planted. The rich wetlands soils produced high yields for farmers—but at the expense of fish and wildlife habitat, flood and erosion control, and water quality. Since 1950, approximately 9 million acres of wetlands have been destroyed in order to increase agricultural production.

The so-called "swampbuster" provisions (16 U.S.C. 3801, 3821–23) of the Food Security Act of 1985, however, may arrest such destruction by removing some of the incentives to convert wetlands to farmland. The act does not prohibit conversions; farmers can still convert wetlands after they obtain a Clean Water Act Section 404 permit from the Corps, but they will be ineligible, with some exceptions, for all U.S. Department of Agriculture (USDA) benefits such as crop price supports, disaster payments, crop insurance, and Farmers Home Administration loans. Such farmers would lose benefits for all crops produced, not just those grown on converted wetlands. Since agricultural conversions represent the greatest source of wetlands losses, these provisions, if vigorously enforced by USDA, could have a tremendous impact on reducing wetlands losses. In 1989, USDA published statistics showing that over 400 farmers lost over $1 million in benefits because of penalties imposed under swampbuster. The National

FIGURE 1.6
EXAMPLES OF WETLANDS LOSSES IN VARIOUS STATES

State	Original Wetlands (Acres)	Today's Wetlands (Acres)	Percent of Wetlands Lost
Iowa's Natural Marshes	2,333,000	26,470	99
California	5,000,000	450,000	91
Nebraska's Rainwater Basin	94,000	8,460	91
Mississippi Alluvial Plain	24,000,000	5,200,000	78
Michigan	11,200,000	3,200,000	71
North Dakota	5,000,000	2,000,000	60
Minnesota	18,400,000	8,700,000	53
Louisiana's Forested Wetlands	11,300,000	5,635,000	50
Connecticut's Coastal Marshes	30,000	15,000	50
North Carolina's Pocosins	2,500,000	1,503,000	40
South Dakota	2,000,000	1,300,000	35
Wisconsin	10,000,000	6,750,000	32

Source: Ralph W. Tiner, Jr., U.S. Fish and Wildlife Service, *Wetlands of the United States: Current Status and Recent Trends*, National Wetlands Inventory Project (Washington, D.C.: U.S. Department of the Interior, Fish and Wildlife Service, March 1984), p. 34.

FIGURE 1.7
EXTENT OF WETLANDS BY TYPE IN CONTERMINOUS 48 STATES

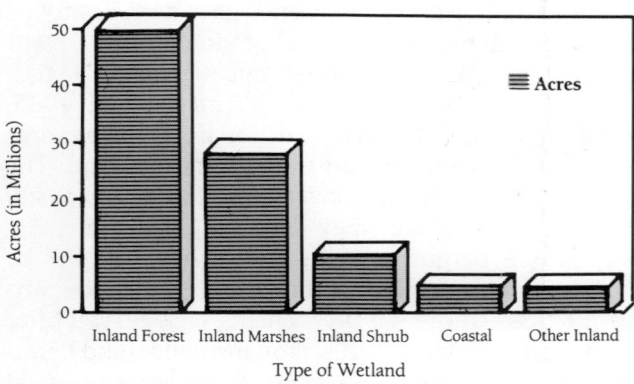

Source: Adapted from U.S. Environmental Protection Agency, *America's Wetlands: Our Vital Link Between Land and Water* (Washington, D.C.: author, February 1988), p. 6.

Wildlife Federation estimates, however, that only about 25 farmers have lost benefits and that the total amount forfeited was about $125,000.[17]

Many agricultural practices, however, such as construction or maintenance of irrigation ponds or ditches, are exempt from the Section 404 permit requirements. And it may still be economical for farmers to convert wetlands to grow crops that do not receive federal subsidies.[18]

The federal government has come a long way since it first encouraged wetlands conversion back in the mid-19th century with the Swamp Lands Acts of 1849, 1850, and 1860, which gave over 60 million acres of wetlands to 15 states for conversion to farmland. Today, the federal government operates a complex array of programs to regulate and protect wetlands. These programs often conflict and are sometimes controversial. In addition, states have taken on a greater role in wetlands regulation, adopting their own wetlands laws and, in one case, assuming administration of the federal wetlands program. The next two chapters provide an overview of the mixed bag of federal and state wetlands regulations.

Notes

1. Ralph W. Tiner, Jr., U.S. Fish and Wildlife Service, *Wetlands of the United States: Current Status and Recent Trends*, National Wetlands Inventory Project (Washington, D.C.: U.S. Department of the Interior, Fish and Wildlife Service, March 1984), p. 6.
2. See Paul H. Zedler, "California's Vernal Pools," *National Wetlands Newsletter*, May–June 1989, p. 3.
3. See the report in the *Washington Post*, November 16, 1988, p. 2.
4. See J. Scott Feierabend and John M. Zelazny, *Status Report on Our Nation's Wetlands* (Washington, D.C.: National Wildlife Federation, October 1987), p. 7.
5. See Edward Maltby, *Waterlogged Wealth: Why Waste the World's Wet Places* (London and Washington, D.C.: International Institute for Environment and Development, 1986), p. 88.
6. See Feierabend and Zelazny, *Status Report on Our Nation's Wetlands*, p. 7.
7. See Donald Hammer and Robert Bastian, "Wetland Ecosystems—Natural Water Purifiers?," unpublished report of the U.S. Environmental Protection Agency, Washington, D.C., and the Tennessee Valley Authority, Knoxville, Tennessee, 1988.
8. Hammer and Bastian, "Wetland Ecosystems," p. 6.
9. Douglass B. Richter, "Genesis and Cleansing in the Wetlands," *The New American Land Magazine*, September/October 1987, p. 27.
10. As reported in the *Washington Post*, November 16, 1988, p. 2.
11. See the booklet by the U.S Environmental Protection Agency, *America's Wetlands: Our Vital Link Between Land and Water* (Washington, D.C.: author, February 1988), p. 5.
12. Tiner, *Wetlands of the United States*, p. 29.
13. John Kosowatz, "Wetlands Establish Their Worth," *Engineering News Record*, October 1987, p. 30.
14. Tiner, *Wetlands of the United States*, pp. 31–36.
15. See Ralph W. Tiner, Jr., U.S. Fish and Wildlife Service, *Mid-Atlantic Wetlands: A Disappearing Natural Treasure*, National Wetlands Inventory Project (Newton Corner, Massachusetts: a Joint Publication of the U.S. Environmental Protection Agency and the U.S. Fish and Wildlife Service, June 1987), p. 18.
16. See Marc Reisner, "The Emerald Desert," *Greenpeace*, July/August 1989, p. 10.
17. See the *Washington Post*, December 6, 1989, p. 3.
18. The Tax Reform Act of 1986 (Pub. L. 99–514. 100 Stat. 2085) prohibits farmers from deducting, from the special agricultural expensing provision, expenses for draining or filling wetlands.

CHAPTER 2

FEDERAL WETLANDS REGULATION

Federal regulation of discharges into the nation's waters goes back a long way. Since the Rivers and Harbors Act was enacted in 1899, the U.S. Army Corps of Engineers has been responsible for keeping the nation's navigable waters open and flowing smoothly. Section 10 of the act prohibits dredging or discharging material in navigable waters of the United States without a permit from the Corps, whose jurisdiction includes waters that are, have been, or might be navigable for interstate commerce.

For many years, the primary thrust of the Corps' regulatory program was to maintain navigability for interstate commerce. New laws and judicial decisions in the late 1960s and 1970s, however, significantly broadened the Corps' jurisdiction from just navigable waters to *all* U.S. waters, including wetlands. The first major expansion in the Corps' role came in 1968, when, in response to growing national concern regarding protection of the environment and pressure from FWS, the Corps revised its permit review process under the Rivers and Harbors Act to include consideration of environmental values. The Corps, which was best known and equipped as a master dambuilder, canal digger, and keeper of the nation's shipping channels and harbors, suddenly was required to make decisions based on a project's impact not just on navigation, but also on fish and wildlife, pollution, ecology, aesthetics, and other factors in the general public interest. This became known as the "public interest review," in which the Corps balanced a project's reasonably foreseeable adverse impacts (habitat destruction, pollution) with its

> *Men with the muckrake are often indispensable to the well-being of society, but only if they know when to stop raking the muck.*
>
> Theodore Roosevelt[1]

positive impacts (economic development, jobs, tax revenue).

The following year, 1969, the National Environmental Policy Act was enacted, which further expanded the scope of the Corps' public interest review. And in 1972, the Federal Water Pollution Control Act, later amended and renamed the Clean Water Act, was enacted. This act was primarily intended to begin the arduous process of making the nation's dirty waters "fishable and swimmable" again, but it has been used to stretch the Corps' jurisdiction from navigable waters to wetlands. Thus, after 70 years of administering a fairly straightforward and uncontroversial program, the Corps was now expected to balance the often conflicting objectives of environmental protection and economic development and to consider the effect of development not just on navigable waters but also on wetlands.

Various federal laws can restrict development in wetlands. But it is the Clean Water Act, and, in particular, Section 404 of the act, that has had the greatest impact on development in wetlands and that has generated most conflicts. Under Section 404, a permit is required from the Corps before dredge or fill materials can be discharged into waters of the United States, including wetlands. As construed by EPA, the Corps, and the courts, the definition of "discharge" is broad and includes filling any U.S. waters for any type of development. This chapter first briefly discusses several pertinent federal laws affecting development in wetlands and then offers a more lengthy analysis of the Clean Water Act and its implications for development.

FEDERAL LAWS AFFECTING WETLANDS

The National Environmental Policy Act of 1969 (42 U.S.C. 4321 et seq.)

In December 1969, Congress passed the National Environmental Policy Act (NEPA) and ushered in what is often called "the environmental decade." The act reflected growing concern throughout the United States that unfettered economic growth was spoiling the water, air, and land on which all life depends. NEPA was enacted to reconcile conflicts between economic growth and environmental protection. It directs all federal agencies to consider the impacts of "major federal actions" on the environment. NEPA does not prohibit development in environmentally sensitive areas, but requires all federal agencies, in making decisions about federal or federally permitted projects, including private projects requiring federal permits, to consider environmental impacts of a proposed federal action. Section 102(2) of NEPA states that "all agencies of the federal government shall . . . insure that presently unquantified environmental amenities and values be given appropriate consideration in decision-making along with economic and technical considerations. . . " (NEPA 42 U.S.C. 4321–4347).

NEPA created the Council of Environmental Quality (CEQ) as an agency in the Executive Office of the President. CEQ was originally created to coordinate federal compliance with NEPA, but since Congress did not grant it authority to adopt regulations, CEQ exercised only an advisory role. In 1978, however, it was granted authority to issue regulations that provide an interpretation of NEPA and that establish uniform procedures for preparing environmental impact statements and environmental assessments.

The environmental impact statement is the heart of NEPA. Under Section 102(2)(C) of the act, federal agencies must prepare a detailed statement, known as an EIS, for "major federal actions significantly affecting the quality of the human environment." The EIS must include a statement of:

- the environmental impact of the proposed action;
- any adverse environmental effects that cannot be avoided should the proposal be implemented; and
- alternatives to the proposed action.

In practice an EIS involves extensive environmental analysis, evaluation of all reasonable and practicable alternatives to the proposed project, considerable interagency review, takes two to three years to complete at considerable expense, and is normally several inches thick. In one extreme case, an impact statement consisted of 24 volumes, each with between 300 and 500 pages. Although EISs are usually required for "major" projects only, they are triggered not just by the size of the project, but also by the value of the resources affected and the magnitude of the controversy.

An environmental assessment, rather than an EIS, is sufficient for most projects that fall under NEPA. It is usually a short document of five to 10 pages that can be completed in a few days or weeks. Like a mini-EIS, an assessment briefly describes the purpose of the project and the likely environmental impacts, offers an analysis of alternatives, and indicates whether or not the impacts will be significant. In most cases, an assessment results in what is called a Finding of No Significant Impact (FONSI), but in a few cases the impacts will be significant enough to warrant a full-blown EIS.

The EIS process can either harm or help a developer's plans. In some cases, the process can delay project approval, cost hundreds of thousands of dollars, and consume the equivalent of a small forest in paper. Or worse, after all the time and expense of preparing an EIS, a project proposal may be rejected. For highly controversial projects, the EIS can become the lightning rod for opposition. It will be discredited if it fails to anticipate and address controversial impacts and reasonable alternatives, or if it was evidently conceived merely to justify a project.

On the other hand, the EIS process can help local, state, and federal agencies to focus on environmental issues and cooperate in the analysis of real problems. In many cases it has helped project proponents channel criticism into constructive paths and useful studies. It has also improved decision-making efficiency by establishing firm schedules for organizing and analyzing data necessary under many different statutes.

CEQ regulations specify what impacts a federal agency should analyze in an EIS, such as direct and indirect effects of a proposed action, possible conflicts between federal, regional, state, and local land use plans, policies and controls, and the environmental effects of alternatives to the proposed action. A federal agency may, though, have difficulty determining the appropriate "scope" of its NEPA analysis. For example, if a developer proposes to build a large industrial complex on uplands and a service road through several acres of adjacent wetlands, it will need a 404 permit from the Corps. Should the Corps, in complying with NEPA, evaluate the environmental impacts of both the upland industrial complex and its service road through wetlands or the road only? Should it evaluate whether the adverse impacts could be reduced by constructing the entire plant at a different location?

Recently, the Corps' NEPA analysis was significantly limited. Until 1988, the Corps' NEPA policy had been to analyze the direct and indirect impacts of a project where a federal permit was required. In 1984, the Corps proposed to change its NEPA procedures in order to avoid situations in which the "federal tail wags the non-federal dog."[2] The new procedures would confine its analysis of a project's environmental impacts to that part of a project requiring a Corps permit. In the above example, for instance, since a 404 permit was required for the road but not for the industrial complex, the Corps would limit its NEPA analysis to the impacts of the road. EPA objected to the Corps' proposed changes and, in 1985, referred the matter to CEQ, which is supposed to resolve such interagency squabbles. In 1987, CEQ finally settled the longstanding dispute and largely accepted the Corps' proposed changes. In February 1988, the Corps issued rules to "clarify and streamline" its procedures for assessing the environmental impacts of projects under NEPA. The new rules significantly narrow the Corps' scope of analysis and limit the reach of federal interest in the environmental impacts of private projects, including projects in wetlands.

Section 404 permit applications are subject to the environmental impact provision of NEPA, but the Corps has estimated that less than 0.5 percent of applications cover projects that, because of their likely impacts on the environment, will require an EIS.[3]

The Clean Water Act of 1977 (33 U.S.C. Sections 1251–1376)[4]

The Clean Water Act (CWA) bolstered the continuing expansion of the Corps' role as wetlands protector. Section 404 of the act prohibits the discharge of dredge or fill material into "navigable waters," defined as "waters of the U.S.," without a permit from the Corps. The

---

OTHER FEDERAL LAWS AFFECTING WETLANDS

The Coastal Zone Management Act of 1972 (16 U.S.C. Sections 1451 et seq.)

Provides financial incentives for states to adopt federally approved coastal zone management programs to protect coastal resources, which include beaches, barrier islands, barrier reefs, dunes, and wetlands. Federal actions, such as offshore oil leasing, must conform with a federally approved state program. If not, the state may "veto" the federal action. This is the so-called "consistency requirement," which has been the focus of considerable debate and litigation between the states and the federal government.

Approved state programs must: 1) delineate the coastal zone boundary; 2) indicate which activities are permissible within the defined coastal zone; 3) inventory special resource areas requiring protection; 4) establish a policy framework to guide decisions about appropriate resource use and protection; and 5) include sufficient legal authority to implement the program.

About 24 of the 30 coastal states, including the Great Lake states, have federally approved coastal zone management programs.

The Endangered Species Act of 1973 (16 U.S.C. Sections 1531 et seq.)

Enacted to protect rare plants and animals, such as the California Condor, that are in danger of becoming extinct. The act requires federal agencies, in consultation with the U.S. Fish and Wildlife Service and the National Marine Fisheries Service, to ensure that any action authorized will not jeopardize endangered or threatened species directly, nor hurt or destroy their habitat, including wetlands. It also prohibits any person from "taking" an endangered species. Taking includes hunting, trapping, harming, or harassing such species.

The National Flood Insurance Act of 1968 (42 U.S.C. Sections 4001–4128)

Provides financial incentives for communities to adopt federally approved floodplain management programs. Administered by the Federal Emergency Management Agency (FEMA), the program utilizes a financial carrot and stick to coax communities into adopting programs that will ultimately reduce the loss of lives and property from floods. For communities with approved programs, the federal government provides subsidized flood insurance to those who own property in the floodplain (the carrot). Communities that do not participate in a program to regulate future floodplain uses are ineligible for federal disaster assistance (the stick). In general, the programs apply to new and rebuilt construction in floodplains, and usually include restrictions on the type and location of development. Although not its primary focus, the program covers development in wetlands, since nearly all coastal and most inland wetlands occur in floodplains.

The Coastal Barrier Resources Act of 1982 (16 U.S.C. Sections 3501–3510)

The act restricts, and in some cases eliminates, federal subsidies for building on undeveloped coastal barriers. The act does not prohibit development on coastal barriers, but it does prohibit federal expenditures and financial assistance, such as federal flood insurance, for such development.

Fish and Wildlife Coordination Act of 1934, amended 1946, 1958, 1977 (U.S.C. 661–667e)

The act requires the Corps to consider the comments of federal and state fish and wildlife agencies, such as the U.S. Fish and Wildlife Service or the National Marine Fisheries Service, before issuing a Section 404 permit.

geographic scope of Section 404 is much broader than Section 10 of the Rivers and Harbors Act, encompassing more than just traditionally navigable waters.

Section 404 of the CWA has three main parts:

- **Section 404(a)** authorizes the Corps to issue permits for filling navigable waters, which includes wetlands. The act states that the "Secretary may issue permits, after notice and opportunity for public hearings, for the discharge of dredged or fill material into the navigable waters at specified disposal sites."[5] The act gave the Corps authority to issue permits, but no guidelines to evaluate them. So the Corps relies heavily on its public interest review (discussed on pp. 30–32) to evaluate permits.

- **Section 404(b)** requires, in essence, that the Corps issue permits in accordance with guidelines developed by EPA—the so-called "b-1 guidelines." The guidelines state that, among other things, "no discharge of dredged or fill material shall be permitted if there is a *practicable alternative* to the proposed discharge which would have less adverse impact on the aquatic ecosystem. . . ."[6] In addition, "no discharge of dredged or fill material shall be permitted which will cause or contribute to *significant degradation* of the waters of the U.S."[7]

- **Section 404(c)** authorizes EPA to veto a decision by the Corps to issue a permit to fill in a wetland. It states: "The Administrator is authorized to . . . deny or restrict the use of any defined area for specification . . . as a disposal site whenever he determines . . . that the discharge . . . will have unacceptable adverse effect on municipal water supplies, shellfish beds and fishery areas (including spawning and breeding areas), wildlife, or recreational areas."[8]

Lesser known but still significant parts of Section 404 are subsections (e), (f), and (g).

- **Section 404(e)** authorizes the Corps to issue general permits on a state, regional, or nationwide basis for certain categories of activities in wetlands that are "similar in nature, [and] will cause only minimal adverse effect to the environment."

- **Section 404(f)** exempts certain activities from the permit requirements, such as "normal farming, silviculture, and ranching activities, . . . minor drainage, harvesting for the production of food, fiber and forest products, or upland soil and water conservation practices." Such activities, however, must be part of an established farming, silviculture, or ranching operation; otherwise, they are not exempt. For example, farmers may not drain and fill a wetland that has never been farmed without first obtaining a permit from the Corps. In one case, Cumberland Farms of Connecticut, Inc., converted a wetland to farmland and was later fined and ordered by the Corps to restore the altered wetland to its previous condition. Cumberland Farms began converting over 1,000 acres of a 2,000-acre cedar swamp to farmland in 1972. In 1984, the Corps issued a cease and desist order and required that Cumberland restore the converted wetland. Cumberland, which had not applied for a 404 permit, challenged the Corps' order. The dispute was settled in 1987 when the First Circuit Court of Appeals ruled that Cumberland Farms' activities were neither exempt from the permit requirements nor covered by a nationwide permit. The court ordered Cumberland Farms to pay a fine of $540,000 (although most of the fine will be waived when the wetland is restored) and to restore the swamp to its 1977 condition—the year in which the swamp first came under the Corps' jurisdiction (*United States v. Cumberland Farms of Connecticut, Inc.*, 826 F2d 1151 (1st Cir. August 1987)).

- **Section 404(g)** authorizes states to assume the permit program from the Corps (except in the case of coastal waters), provided their program is approved by EPA.

One of the more unusual aspects of Section 404 of the CWA is that, although EPA administers most federal environmental laws including the CWA, it does not administer Section 404. When Congress was drafting that section, it decided that the Corps was best suited to run the program, given its long history of protecting navigable waters and administering a permit program under the Rivers and Harbors Act. Congress was also mindful, however, that the Corps—an agency that has been dredging and filling wetlands for years—might be too prodevelopment and thus gave EPA oversight of the 404 permit program. In this forced marriage, the Corps issues permits, but it must follow guidelines set and monitored by EPA. In addition, following amendments to the Fish and Wildlife Coordination Act, the Corps must give "full consideration" to comments by the FWS and the National Marine Fisheries Service (NMFS) when reviewing permit applications. A General Accounting Office (GAO) report found, however, that although the Corps generally considers the comments from resource agencies, it frequently disregards their advice. GAO estimated that the Corps issued permits over the objections of the resource agencies in 37 percent of the 111 cases GAO reviewed.[9]

If, after considering all the comments and performing the public interest balancing act, the Corps decides to issue a permit, EPA, under Section 404(c), can still "veto" that decision if the agency finds that the proposed project will have unacceptable adverse impacts on the environment. Since the program started, the agency has issued fewer than a dozen vetoes. Probably its most famous veto was in 1986, when EPA overrode the Corps' decision to issue to The Pyramid Companies a permit to fill 32 acres of a 50-acre wetland in Attleboro, Massachusetts, for a shopping mall (see feature box on pp. 26–27). Thus, while Congress sought to balance the administrative efficiency of the Corps with the environmental protection mission of EPA, it sowed the seeds of interagency squabbling that continues today.

The Corps initially interpreted the CWA to cover only traditionally navigable waters. But, in 1975, the Natural Resources Defense Council (NRDC) challenged the Corps' narrow interpretation of the act (*NRDC* v. *Calloway*, 392 F. Supp. 685 D.D.C. 1975). The District Court for the District of Columbia agreed with NRDC's contention that the term "waters of the U.S." was to be interpreted broadly. Based on this decision, the Corps adopted new regulations in 1975 that redefined "waters of the U.S." to include navigable waters and their tributaries, interstate waters and their tributaries, nonnavigable intrastate waters whose use or misuse could affect interstate commerce, and all freshwater wetlands adjacent to other waters protected by the statute. The NRDC decision vastly expanded the geographical scope of the Section 404 program.

The Corps now defines wetlands as "those areas that are inundated or saturated by surface or groundwater at a frequency and duration sufficient to support, and that under normal circumstances do support, a prevalence of vegetation typically adapted for life in saturated soil conditions" (33 CFR Part 328.3 (b)). Wetlands include marshes, bogs, swamps, and similar areas. This definition was upheld by unanimous ruling of the U.S. Supreme Court in 1985 (*U.S.* v. *Riverside Bayview Homes, Inc.* 106 S.Ct. 455 (1985)).

Twelve years ago, Riverside Bayview Homes, Inc., began filling several acres of its 80-acre property near Lake St. Clair in Macomb County, Michigan. The Corps issued a "cease and desist" order, which was followed by a district court preliminary injunction against further filling. The Corps claimed that Riverside Bayview needed a permit to fill because, based on the Corps' new definition, the property was classified as an adjacent wetland. Riverside Bayview argued that the presence of aquatic vegetation on its property was not due to frequent flooding by nearby navigable waters, and therefore it did not fall under the Corps' jurisdiction. The trial court upheld the Corps' jurisdictional claim, but this decision was reversed by the court of appeals. The U.S. Supreme Court, however, rejected the appeals court's claim that only wetlands that are inundated or frequently flooded by an adjacent body of water fall under the Corps' jurisdiction. In reviewing the legislative history of the CWA, the Supreme Court cited Congress's interest in protecting wetlands and noted Congress's rejection of attempts to narrow the definition of navigable waters during the debate over the 1977 CWA amendments. The Court deferred to the Corps' expertise and upheld its definition and jurisdiction over wetlands adjacent to waters of the United States.

The *Riverside Bayview* decision significantly expanded the Corps' jurisdiction over wetlands. Now wetlands do not have to be wet, connected to, or flooded by navigable waters to fall within the Corps' purview; they must only be saturated frequently enough to support wetlands vegetation and have some connection to interstate commerce. The Supreme Court required only that wetlands be hydrologically connected with navigable waters. Unfortunately, however, the Supreme Court ducked the issue of whether or not isolated wetlands—wetlands that are not adjacent or connected to waters of the United States—fall within the Corps' jurisdiction. But, chances are, if a duck lands in it, it is a wetland and is probably covered by Section 404. As long as there is some connection to interstate commerce, which the Corps is supposed to protect, then the waters fall within the Corps' jurisdiction. For the most part, it does not take much to establish such a connection. For example, if a hunter in Texas shoots a duck that bred or fed in an isolated wetland in North Dakota, that duck is considered an item of interstate commerce and the isolated wetland is regulated by the Corps.

From its very beginning in 1972, the 404 permit program has been a constant source of litigation and controversy. State and local governments often view the program as an unwarranted intrusion on local land use controls. Developers find the program overly burdensome, and the two federal regulators with primary responsibility for administering the program—EPA and the Corps—have fundamentally different interpretations of the act's purpose.

The stated purpose is to "restore and maintain the chemical, physical and biological integrity of the nation's waters."[10] The Corps has long maintained that the act directs it to protect navigation and water quality but not specifically to protect wetlands. Indeed, if one of the goals of the CWA is to protect wetlands, then Congress was rather careless when it crafted the legislation. The act contains so many loopholes and exemptions that most wetlands can be legally destroyed. For example, one can legally drain, dig ditches through, or dig large holes in a wetland without a permit, as long as none of the dirt, mud, or sand is deposited in the wetland.

EPA, in contrast, believes Congress clearly intended that the program be used to protect wetlands; if not, Congress would have explicitly stated otherwise when it amended the CWA in 1977. As a result of EPA's and

SHOOTOUT AT SWEEDENS SWAMP: THE ATTLEBORO MALL CASE

In December 1983, a shopping center developer called The Pyramid Companies bought an 82-acre tract near Interstate 95 at the southern edge of Attleboro, Massachusetts. The property contained a 50-acre red maple swamp known as Sweedens Swamp. Pyramid proposed to fill 32 acres of the swamp in order to build a $40 million, 700,000-square-foot regional shopping mall. The proposed mall would have necessitated the largest permitted fill in Massachusetts.

To mitigate the adverse effects of the fill, Pyramid proposed to create a 36-acre wetland out of an abandoned gravel pit a few miles off site and to excavate an additional 13 acres of wetlands and nine acres of uplands to create additional wetlands on site.

After obtaining all necessary state and local approvals, including a permit—upheld by the state supreme court—under the Massachusetts Wetlands Protection Act, Pyramid applied to the Corps for a Section 404 permit in July 1984.

EPA and other resource agencies objected to the proposed project, and the Corps' New England Division office was prepared to deny the permit on the grounds that

EPA successfully argued that Attleboro Mall should be located elsewhere to protect this red maple swamp.

alternative, nonwetlands sites were available. The Corps' headquarters in Washington, D.C., however, felt that the division office was interpreting the 404(b)(1) guidelines too narrowly and ordered it to issue the permit.

In June 1985, the Corps issued a permit contingent upon the success of the proposed off-site mitigation, which, the Corps believed, would reduce the overall negative impacts to zero. In May 1986, however, EPA invoked its 404(c) authority and vetoed the permit, because, in its opinion, a shopping center is not water-dependent and therefore could be located in an upland area, and an alternative upland site was available to Pyramid. The agency found that the proposed mitigation was irrelevant and inappropriate since the fill was avoidable. In EPA's view, mitigation is appropriate only when wetlands losses are unavoidable.

The 404(b)(1) guidelines presume that practicable alternatives exist for nonwater-dependent activities; EPA determined that Pyramid had not overcome this presumption. Pyramid filed suit in Federal District Court to overturn EPA's veto on the grounds that there were no economically viable alternatives to the Sweedens Swamp site. Two of the key issues addressed by the court were, one, whether EPA has the authority independently to apply the practicable alternatives test of the 404(b)(1) guidelines as part of its 404(c) analysis, and, two, exactly when is an alternative site considered "available": when the developer first enters the market, as EPA contended; or when the developer applies for a permit. In October 1987, the court upheld EPA's veto and ruled that EPA does have the authority to conduct its own alternatives analysis. The court also found that there was a less environmentally damaging site (which Pyramid had earlier turned down) at the time Pyramid entered the market (*Bersani v. EPA*, 674 F.Supp. 405 (N.D. N.Y. 1987)).

On appeal, Pyramid challenged what it called EPA's "market entry" theory and argued that there were no prac-

LOCATION OF PROPOSED ATTLEBORO MALL

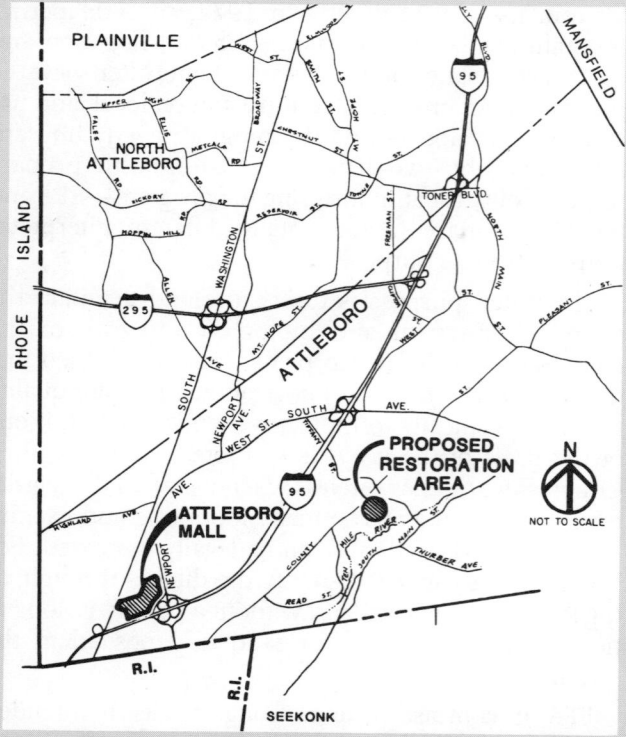

Source: Prepared by BSC Engineering, Boston, July 1985.

A few readily accessible areas of Sweedens Swamp near Attleboro, Massachusetts, became a popular dump site with local residents.

ticable alternative sites at the time it applied for a permit. EPA, however, asserted that Pyramid began searching for a site in the spring of 1983. The difference is crucial to the case because sometime between when Pyramid began looking for a site and when it applied for a permit, a competing developer purchased an option to buy an upland site in North Attleboro. This site was considered the only viable alternative upland site for the mall, but as such it was also a practicable alternative.

In a 2 to 1 decision, the court upheld EPA's market entry theory and held that: one, the market entry theory is consistent with both the regulatory language and past practices; two, EPA's interpretation of the statute is reasonable; and three, EPA's finding is not arbitrary and capricious. Pyramid petitioned the U.S. Supreme Court to hear the case, but in March 1989 the Court declined to review it.

The Attleboro Mall battle, which received national attention, produced some unusual alliances. On one side were groups who opposed the project: EPA, FWS, the New England district of the Corps, local and national environmental groups, and a competing developer. On the other side were The Pyramid Companies, the Washington, D.C., office of the Corps, and the city of Attleboro.

Both sides would admit that Sweedens Swamp is not a pristine wetland; it is an ordinary red maple swamp traversed by I-95, and parts of the swamp have been used as an occasional dump by local residents. But, according to EPA, despite a few alterations and abuses, the swamp provides excellent wildlife habitat, which should not be destroyed by development that can and should occur on uplands. In addition, given the uncertainties surrounding newly created wetlands, particularly manmade freshwater swamps, EPA was unwilling to swap an existing red maple swamp for a gravel pit-cum-wetland. The agency chose to take a stand on this particular wetland because it feared that this project could set a dangerous precedent of allowing nonwater-dependent development to occur in wetlands.

The Attleboro Mall decision underscored the institutional problems created by EPA's and the Corps' competing missions and divided, but poorly defined, responsibilities. For years, the two bodies have been battling over their respective roles and interpretations of Section 404 of the CWA. Their differences are often played out at the expense of the private sector, which must guess at the outcome while the agencies struggle to reach a decision. The Attleboro decision clearly illustrates the sometimes unpredictable, time-consuming, and litigious nature of the 404 process. The case, however, also served to endorse EPA's broad authority not only to second guess a permit applicant's selection of a development site, but also to oversee and, if necessary, to veto a Corps permit. The agency can confidently assert its authority to reevaluate the Corps' alternatives test to ensure that the Corps strictly adhered to the 404(b)(1) guidelines.

Unfortunately, the case did not give developers clear guidance on exactly when they must consider practicable alternatives or how extensively they must search to find an alternative site: five miles, 50 miles, 100 miles? In his dissent, Judge Pratt stated that the market entry theory will "inject exquisite vagueness" into the permitting process. For example, when does a developer enter the market? When it first contemplates a development in the area? When it first instructs its staff to research possible sites? When it contacts a real estate broker? Or when it makes its first offer to purchase? Without answers to such questions, developers can never be certain exactly when the clock starts. But developers can be sure of one thing: from now on, it will be much harder to get a permit to fill wetlands for nonwater-dependent uses.

Despite laws to protect them, thousands of acres of wetlands succumb to residential development each year.

the Corps' inconsistent interpretations, developers are often caught in the crossfire of policy battles between the two agencies.

Through a series of court cases and regulatory changes, more so than through congressional action, Section 404 has evolved from a water-quality statute to a wetlands protection statute. Similarly, the Corps has evolved from a staunch protector of navigable waters to a reluctant wetlands protector. Recently, the Corps has shown greater willingness to assert its authority to protect the nation's wetlands, and both EPA and the Corps are beginning to work together on a number of fronts. For example, they recently reached an agreement on a mitigation policy and have been evaluating alternatives to the case-by-case permitting process. In addition, in 1989, EPA and the Corps, along with FWS and the Soil Conservation Service, developed a joint manual for delineating wetlands in the field.

Prior to adoption of the joint manual, the four agencies each had their own method of delineating wetlands. In general, the Corps was conservative in its delineations, FWS was very liberal and typically included some wet places that the Corps would not, and both EPA and the Soil Conservation Service fell somewhere in between. (EPA and the Corps utilized separate delineation manuals, although they had adopted a single definition of wetlands.) Moreover, considerable variation occurred within the agencies themselves in delineating wetlands. Understandably, the different wetlands definitions and manuals led to confusion in the regulated community.

The joint manual relies on three criteria to delineate wetlands: hydrophytic vegetation, hydric soils, and wetlands hydrology. It reflects a compromise among the agencies and should clear up any confusion by providing a single, consistent approach for identifying and delineating wetlands. Nonetheless, the manual has brought some painful groans from the development community who claim that the new approach slightly expands the Corps' jurisdiction into some lands not previously considered wetlands. Indeed, a Maryland developer has sued EPA and the Corps on the grounds that the manual is actually a rule implemented by the agencies without complying with the Administrative Procedure Act (see *Mulberry Hill Development Corporation* v. *U.S.*, Civil Action No. PN89-2639).

PERMITTING UNDER SECTION 404 OF THE CLEAN WATER ACT

"We are getting to the point of adopting a Nancy Reagan approach to wetland fills: 'just say no'."

Official at EPA Region X

Just about everyone who proposes to fill in a wetland must first apply for a permit from one of the Corps' 36 district offices. Applying for a permit from the Corps is not something anyone looks forward to. It's a bit like going to the dentist: at the very least it will be uncomfortable, at worst it will cause agony at great expense. For small, noncontroversial projects, the process is fairly straightforward and will, under normal circumstances, take only a few months to complete. For larger, more complex projects, however, getting a permit can be a long, arduous process.

Not all activities in wetlands require a 404 permit and some require only a "general permit"—a kind of generic permit that grants blanket authorization for certain types of fill in certain size wetlands. The Corps may issue general permits on a state, regional, or national basis for activities that are similar in nature and will cause only minimal individual and cumulative adverse impacts. For example, discharges for bridge construction, backfill for utility lines, surveying, and minor road crossings usually qualify for a general permit. General permits issued nationwide are called, appropriately enough, "nationwide permits." The nationwide permit most often used for development is nationwide permit number 26, which applies to certain isolated wetlands and wetlands above a nontidal river or stream with an average annual flow of five cubic feet per second or less. Nationwide permit number 26 is usually issued automatically for wetlands fills of one acre or less, while those who want to fill between one and 10 acres must first notify the Corps. For controversial projects, the district engineer may require an individual permit instead. General permits, which are issued for five-year periods, apply only in states that certify that the activities authorized by such permits comply with state water-quality requirements. Some states, such as New Jersey, do not recognize such permits.

The Corps receives about 14,000 applications for individual dredge and fill permits each year. Of these, about 10,000 are granted, usually with conditions to reduce environmental impacts. Only about 500 are denied. The rest are either withdrawn or are covered by general permits. Most applications are fairly routine, and relatively few projects stir up enough controversy to make the headlines.

The permit process consists of three basic steps: preapplication consultation; formal project review; and decision making. All three usually occur at one of the Corps' many district offices. The steps are described below and are shown graphically in Figure 2.1.

Preapplication Consultation

The preapplication consultation is an informal meeting between a district office of the Corps and the applicant to discuss the permit requirements, identify potential problem areas, and to determine, on a preliminary basis, whether an EIS will be required. In this frank and open consultation, the Corps advises the applicant of the factors that the Corps must consider under the CWA Section 404(b)(1) guidelines and in its public interest review. The Corps may determine that only a general permit is required, or, occasionally, that a permit is not needed—for example, if the project is not in U.S. waters.

The consultation is optional and is recommended only for large or potentially controversial projects. Besides the Corps, other federal agencies, state agencies, and local affected groups sometimes participate in the consultation. This enables such agencies or groups to air their concerns upfront, and will give an applicant an opportunity to address them early in the permitting process, rather than later when the stakes are higher. Most applicants do not bother with a preapplication consultation. In fact, the district offices of the Corps are often so swamped with applications that they do not have time for consultations, and the applicant must submit a "blind" application. A consultant in Oregon complained that the Corps will no longer meet with developers until it has received a formal application.

Formal Project Review

Applications must include, among other things:
- a detailed description of the proposed project, including maps, drawings, and sketches that can be used for a public notice;
- the purpose of the project;
- scheduling of construction; and

FIGURE 2.1

THE U.S. ARMY CORPS OF ENGINEERS' PERMITTING PROCESS

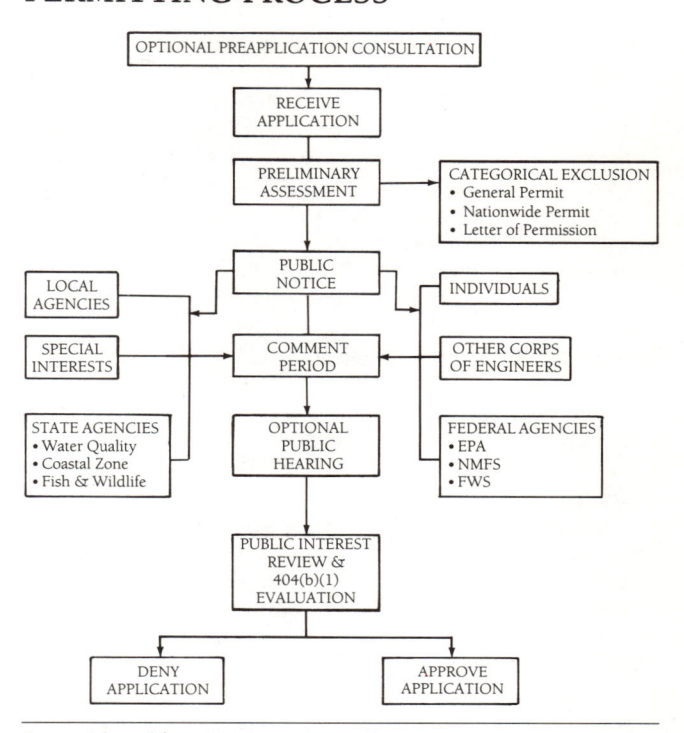

```
OPTIONAL PREAPPLICATION CONSULTATION
          ↓
   RECEIVE APPLICATION
          ↓
   PRELIMINARY ASSESSMENT → CATEGORICAL EXCLUSION
                              • General Permit
                              • Nationwide Permit
                              • Letter of Permission
          ↓
LOCAL AGENCIES → PUBLIC NOTICE ← INDIVIDUALS
          ↓
SPECIAL INTERESTS → COMMENT PERIOD ← OTHER CORPS OF ENGINEERS
          ↓
STATE AGENCIES → OPTIONAL PUBLIC HEARING ← FEDERAL AGENCIES
• Water Quality                              • EPA
• Coastal Zone                               • NMFS
• Fish & Wildlife                            • FWS
          ↓
PUBLIC INTEREST REVIEW & 404(b)(1) EVALUATION
          ↓                    ↓
   DENY APPLICATION      APPROVE APPLICATION
```

Source: Adapted from U.S. Army Corps of Engineers.

- the type of discharge, (sand, gravel, dirt, etc.).

Incomplete applications will be returned, often delaying the process for several weeks.

Within 15 days after the Corps receives a completed application, the district engineer issues a public notice. The notice is sent to EPA, FWS, NMFS, state and local agencies, special interest groups, and other interested agencies, organizations, and individuals. After the public notice is issued, the Corps usually allows 15 to 30 days, depending on the nature of the activity, for public comments. The Corps' goal is to reach a decision within 60 days after the public notice is issued, but it normally takes closer to a year for controversial projects.

When reviewing a 404 permit application, the district engineer must, as required by NEPA, determine whether an EIS or only an environmental assessment is needed. EISs are rare for Section 404 projects—the Corps requires only about 15 per year. Whether or not an EIS is required, certain applications may involve additional interagency or intergovernmental consultations. For example, the Corps may consult with federal and state agencies to ensure that the proposed project is consistent with a state coastal zone management program, the Endangered Species Act, or the National Historic Pres-

HOW THE SECTION 404 PERMIT PROCESS CAN SHAPE DEVELOPMENT: THE PRUDENTIAL BUSINESS PARK

In northern New Jersey, not far from New York City, lies a 367-acre wooded site that will soon be the home of the Prudential Business Park. The site contains about 230 acres of forested wetlands and is considered one of the last remnants of an extensive forested wetland that once occupied the area. It is also one of the last remaining large natural areas in that part of the state (Parsippany Troy Hills and Hanover Township), which is why so many people fought to preserve it.

The site is encircled by development: a residential area on the north, I-287 on the east, route 10 on the south, and the existing Prudential Business Campus on the west. Prudential plans on building 2.2 million square feet of commercial office space in a campuslike environment. Only about 16 acres of wetlands will be filled. In exchange, Prudential will leave most of the site open and restore a degraded wetland off site.

Prudential's initial proposal, made in 1982, called for constructing over 4.5 million square feet of office space on 250 acres and would have eliminated over 100 acres of wetlands. The proposal drew immediate criticism from EPA, FWS, environmental groups, and the New Jersey Department of Environmental Protection (DEP), who argued that the proposed project is not water dependent. In response, Prudential reluctantly whittled the project down, first to 3.5 million square feet and ultimately to 2.2 million square feet, and reduced the area to be filled from about 100 to only 16.6 acres. Several revisions and about four years later, Prudential finally obtained its 404 permit in February 1988 and began construction that fall. The project should be completed in about seven to nine years.

Touch and Go Permitting

After a relatively smooth start, the permit process turned into a tortuous struggle. Prudential approached the Corps in 1981 to determine if what was then only a 200-acre parcel fell within the Corps' jurisdiction. The Corps' New York district office initially found that the project would have a minimal impact on the aquatic environment and therefore would be covered by a nationwide permit. But after Prudential purchased an additional 167 acres contiguous to the 200-acre parcel, the Corps decided that an individual permit would be required instead. The Corps reasoned that, while the original 200 acres was primarily forested upland and forested wetland, the extra 167 acres contained mostly wetlands, and thereby significantly increased the impact of Prudential's project on the aquatic environment.

Prudential, which had already invested about $8 million in the project, challenged the Corps' authority to rescind its nationwide permit and also questioned whether the Corps had jurisdiction over the wetlands in the first place, since the wetlands were "artificial" and existed only because a water impoundment had been constructed downstream in the 1940s. In July 1984, the Corps reaffirmed its requirement that Prudential obtain an individual permit for their project. Brushing aside Prudential's challenges, the Corps stated that its jurisdiction is based on current conditions, and not on whether a particular wetland is natural or manmade. In a number of cases, the courts have upheld the Corps' jurisdiction over artificial wetlands. For example, in one case, the plaintiff contended that the wetlands on his property were created by construction of a dam and therefore did not fall within the Corps' jurisdiction. The court held that the Corps' jurisdiction is based on the current state of a site, not its history; the origin of the wetland was irrelevant. (See *Bailey* v. *United States*, 647 F.Supp 44 (D. Idaho 1986); see also *Track 12, Inc.* v. *District Engineer*, 618 F.Supp. 448 (D. Minn. 1985).)

New Jersey's response to Prudential's original proposal was, according to a DEP official, that "the project is no good as it stands." The state laid down a number of conditions that Prudential had to satisfy in order to obtain a water-quality certification. The state maintained a steadfast approach to Prudential, and, in the end, the company met virtually all of the state's demands.

---

ervation Act, before it will issue a permit. States may veto a Corps permit if the state determines that the project will be inconsistent with its coastal zone management program or violate its water-quality standards under Section 401 of the CWA.

Decision Making

After a public notice has been issued, an environmental assessment or EIS has been completed, and the project has been determined to be consistent with state laws, the district engineer will decide if a permit should be issued. This decision will be based on two factors: the public interest review, and the CWA Section 404(b)(1) guidelines. The two procedures are usually performed simultaneously, and the Corps may require some mitigation to meet the requirements of either test.

When the Corps conducts its public interest review, it weighs a project's likely impacts on a wide variety of factors: aesthetics, navigation, cultural values, environmental values, economics, water quality, fish and wildlife values, safety, conservation, and so on. If, on balance, the impacts are negative, then the Corps may deny the permit, require that the applicant modify the design, or

Generous Mitigation

After several revisions, each reducing the project size and the extent of the wetland fill, the final design concentrates most of the project on the upland portion of the site and leaves the majority of the site open and undisturbed. Of the site's 232 acres of wetlands, only 16.6 will be filled—about 10.5 acres for roads, and only about six acres for the buildings. The remaining wetlands, about 215 acres, will be incorporated into a conservation easement and preserved in perpetuity.

In addition to the 16.6 acres, 22 acres will be excavated to control floods and maintain water quality. This area will be graded, planted as wetlands, and included in the conservation easement. Prudential will also construct gabion walls at several locations to detain rainfall on the upland portions of the site and protect the wetlands from stormwater contamination.

In an attempt to prove that wetlands creation was a viable mitigation option, Prudential voluntarily created a seven-acre emergent marsh out of upland within a road access loop just east of I-287. Prudential needed fill dirt for the ramps and decided to create a marsh out of the hole that was left by the excavation. The company planted pickerel-weed, arrowhead, wild rice, and other emergents. According to one of the consultants who worked on the project, the marsh is doing reasonably well. EPA, however, is not impressed. It believes that highway runoff will eventually contaminate and destroy the marsh. According to one EPA official, the marsh is just "a hole in the middle of a highway loop."

Prudential also agreed to purchase a 100-acre parcel of land about five miles away and adjacent to the Great Swamp National Wildlife Refuge. The project's nonwater dependency allowed state regulators to raise significantly the mitigation ante. Prudential was not required by the Corps to purchase the Great Swamp site, but it was strongly urged by the state to make additional concessions. The land contains a mix of forested wetlands and pasture. Most of the pasture was originally wetlands that were cleared and drained years ago. The existing, marginal drainage system will be removed to allow the pasture to revert back to forested wetlands through natural succession. To speed up the process, Prudential will plant wetlands trees and shrubs. Prudential donated the 100-acre site to the New Jersey Natural Lands Trust—a state-sponsored conservancy.

As part of the permit requirements, Prudential must ensure that the mitigation will be at least 80 percent successful—that is, at least 80 percent of the plants in the created or enhanced wetlands must survive for five years; if more than 20 percent die, Prudential must replant and do whatever else is necessary to ensure the continued well-being of the wetland for at least five years.

The mitigation is estimated to cost about $5 million, including $1.8 million for the 100-acre tract near the Great Swamp. Prudential will also spend another $25 million for off-site road improvements.

All in all, the regulators certainly got a good deal. Prudential pared down the amount of fill involved, preserved over 215 acres of wetlands on site and another 67 (from the 100-acre site) acres off site, and achieved an overall ratio of wetlands preserved to wetlands filled of about 17:1. EPA, DEP, and the Corps were all satisfied that the proposed mitigation adequately compensated for the wetland fill. According to one EPA official, the 100-acre wetland "donation" is certainly "better than creating a wetland out of a cornfield."[1] It was not a pretty process, and the project has a long way to go before it is complete, but it illustrates what can be done to avoid or minimize the adverse impacts of developing in wetlands, and how the permit process can substantially alter development plans.

Note:
1. Personal conversation on August 18, 1988, with Dan Motella of the U.S. Environmental Protection Agency, Region II, New York, New York.

---

require mitigation. If positive, then the Corps may grant a permit (providing that the project also complies with the 404(b)(1) guidelines). Until 1982, an applicant had to demonstrate that the project would be "in the public interest" in order to qualify for a permit. In 1982, the Corps decided that as long as a project was "not contrary to the public interest"—an easier standard to meet—a permit could be issued.

The 404(b)(1) guidelines were developed by EPA in conjunction with the Corps. By law, the Corps must follow these guidelines when reviewing permit applications, and it must deny a permit for any proposal that fails to comply with them. The guidelines state, in part, that the Corps should not grant a permit:

- if there is a practicable alternative that would have less adverse impact on the aquatic ecosystem;
- if the project will cause or contribute to significant degradation of the waters of the United States; and
- unless appropriate and practicable steps have been taken that will minimize potential adverse impacts on the aquatic ecosystem.

The practicable alternatives test is the heart of the guidelines. It establishes two presumptions that must be overcome by the applicant: one, that an alternative, nonwetland site is available for nonwater-dependent activities; and, two, that activities in nonwetlands areas will have less adverse environmental impact than activities in wetlands.[11] That is, unlike marinas, developments such as malls, condominiums, and office buildings do not have to be near the water, so the Corps should deny a permit for these types of activities if they involve filling wetlands. The practical alternatives test and the water dependency requirement have been called the Scylla and Charybdis of the permitting process.

One of the goals of the public interest review is to ensure that certain projects that provide public benefits, such as jobs or housing, will not be rejected out of hand because of their adverse environmental impacts. At the same time, the guidelines ensure that less environmentally damaging sites will be considered and that adverse impacts will be minimized. Unfortunately, however, both the public interest review and the guidelines are subjective tests with imprecise standards—a combination that spells trouble. In the guidelines in particular, two phrases frequently cause problems: "practicable alternative" and "significant degradation." These phrases are akin to the vague environmental standards found in NEPA, such as "to the fullest extent possible" and "major federal action." EPA defines practicable as "available and capable of being done after taking into consideration cost, existing technology, and logistics in light of overall project purposes."[12] Significant degradation includes "significantly adverse effects of the discharge . . . on human health or welfare, . . . life stages of aquatic life, . . . [and] aquatic ecosystem diversity, productivity, and stability."[13] Of course, what EPA considers practicable or significant may conflict with an applicant's or the Corps' interpretation; and it can take months or years to settle such differences.

Even after the Corps considers all the comments, conducts its public interest review, determines that the project complies with the guidelines, and decides to issue a permit, the applicant is still not out of the woods. EPA may object if it determines that the project will have an unacceptable impact on the environment. In addition, both the Corps and EPA may agree that a proposed project should include some sort of mitigation, but they may differ on the type or extent of mitigation.

THE MITIGATION DILEMMA

When it comes to mitigation, the Corps, on one hand, and EPA, FWS, and NMFS on the other, often have very different policies. In general, EPA strives to maintain the chemical and biological integrity of U.S. waters; FWS and NMFS aim to protect fish and wildlife habitat; and the Corps seeks to balance competing interests. All,

Aerial photo of Point Mouillee in 1984, looking north towards Detroit. The south end of the confined disposal facility or "island" is nearly filled.

however, have adopted the CEQ definition of mitigation, which states that mitigation includes: a) avoiding the impact altogether by not taking a certain action; b) minimizing impacts by limiting the degree of the action; c) rectifying the impact by repairing, rehabilitating, or restoring the affected environment; d) reducing or eliminating the impact over time by preservation and maintenance operations; and e) compensating for the impact by replacing or providing substitute resources.[14] In short, avoid what you can, minimize where possible, and after that, compensate for the damage done.

EPA, FWS, and NMFS have long been at odds with the Corps over the interpretation of the CEQ definition of mitigation. Unlike the Corps, the three agencies have always asserted that the five elements of mitigation, as defined by CEQ, should be followed step by step. In other words, when considering the appropriate mitigation for a proposed project, the first step should be to avoid the impact altogether—for example, by selecting an alternative, nonwetland, site. After avoidance, the next step should be to minimize the impacts and so on. Only after the first four steps have been exhausted would compensation be allowed.

The Corps, in contrast, has long held that it did not have to follow the CEQ definition in sequence and allowed compensation as the first mitigation measure. It argued that mitigation, in any sequence, could reduce the negative impacts of a project and tip the scale in favor of granting a permit.

In November 1989, EPA and the Corps reached a landmark agreement that, in essence, states that the Corps, when reviewing 404 permit applications, will follow the CEQ mitigation steps in sequence and will strive to achieve EPA's cherished goal of no net loss of wetlands. In December, however, the White House postponed plans to implement the interagency agreement in response to strong opposition from the oil industry and state officials from Alaska who feared that the agreement would have disastrous consequences for oil development along Alaska's North Slope.

One of the reasons for the Corps' proclivity to allow mitigation "out of sequence" is that the Corps itself has been dredging, filling, and creating wetlands for years. During the last 15 years, the Corps has spent over $12 million on wetlands research and has led the nation in developing wetlands habitat.[15] It has built, sometimes inadvertently, more than 100 wetlands and 2,000 islands near U.S. waterways as part of its never-ending dredging of rivers and harbors. One of the largest freshwater wetlands construction efforts ever undertaken is being conducted by the Corps at Point Mouillee Marsh, along the shores of Lake Erie, just below the mouth of the Detroit River. Historically, the approximately 2,600-acre marsh was protected by a barrier beach. But over the past 50 years, unusually high lake levels gradually eroded the beach and exposed the marsh to the destructive forces of Lake Erie. In order to protect the waning marsh from further erosion and to find a final resting place for the dredge spoils from the Detroit and Rouge Rivers, the Corps fabricated a barrier beach along the length of the marsh. The banana-shaped, make-shift beach is about 3.5 miles long and about 1,400 feet wide. It has allowed animal and plant denizens of the marsh to recolonize previously washed-out areas.

One of the Corps' most successful wetlands creation projects is Galliard Island in the lower Mobile Bay in Alabama. The island was built between 1980 and 1981 from material dredged to maintain the bay's shipping channels. The triangular, 1,300-acre island provides both upland and wetlands habitats, and contains a shallow, 700-acre pond. Three broad, gently sloping dikes, maintained by the Corps, protect the pond from erosion. Wind and wave erosion constantly batter all sides; but a combination of stone armoring and salt marsh plantings protects its shoreline. In 1981, the Corps planted cordgrass in the intertidal zone along the northwest and south dikes. It also installed erosion-control mats and low-cost, temporary breakwaters consisting of floating and fixed tires to protect the newly planted seedlings.

Culverts at the south and north ends of the confined disposal facility at Point Mouillee slow down water flow and trap sediments that re-form and nourish the growing marsh.

The island was quickly discovered by shorebirds from the Alabama mainland. In 1981, an estimated 4,000 laughing gulls, black skimmers, and terns were nesting on the island. Since 1984, over 16,000 shorebirds have nested there each year.[16]

The Corps is involved in hundreds of other marsh creation/restoration efforts around the country, with varying degrees of success, as it seeks creative, productive uses for the bargeloads of muck it dredges from the nation's rivers and harbors each year. Based on its experience, the Corps believes that properly designed and constructed artificial wetlands can provide the same values and functions as natural wetlands. But EPA, FWS, and NMFS disagree.

As part of their permit review process, the district and regional offices of the Corps and EPA meet regularly and make ad hoc decisions on specific projects. According to Doug Ehorn, EPA's deputy chief of the Water Quality Branch in Region V, 99.5 percent of all cases are handled routinely with no problems. The few problems that do arise usually concern the amount of mitigation that should be required. In those few cases where EPA and the Corps are at loggerheads, the permit decisions are occasionally elevated from the district to the division engineer, and, if necessary, to the Corps headquarters in Washington, D.C.

Each regional office of EPA and each district office of the Corps operates independently, although general policy is established at their respective headquarters in

View of wetlands inside Point Mouillee.

33

Washington, D.C. No two regional or district offices are alike; permit applicants will not get exactly the same response from any two offices. In some regions, EPA and the Corps have a very good working relationship, and they resolve most of their conflicts amicably. In others, like a bad marriage, the relationship is strained, and the two grind out permit reviews with minimal cooperation and communication. In Southern California, for example, EPA and the Corps reportedly seldom speak to one another. Occasionally, philosophical differences or personality clashes between EPA and Corps officials tie the permit process up in knots. Applicants—who cannot legally proceed with the project until a permit is obtained—are left in never-never land while the agencies try to work out their differences.

Applicants who are denied a permit by the Corps or via an EPA veto of a Corps permit, have essentially four choices: withdraw their application; modify the project and reapply; proceed with their project without a permit and hope they do not get caught; or sue the Corps or EPA. In EPA Region V, for example, about one-third of the applicants who are denied permits withdraw their application, about one-third modify their proposal and reapply, and the remainder are split almost equally between those who proceed without a permit and those who sue.[17]

Enforcement

Both EPA and the Corps have authority to seek criminal, civil, or administrative remedies for unauthorized activities in wetlands. The Corps focuses its enforcement efforts on violations of both Section 10 and Section 404 permits, while EPA, which was given independent enforcement authority under the CWA of 1977, focuses on unauthorized discharges. Enforcement by both agencies, however, has been lax. The agencies simply do not have the resources for a vigorous monitoring and enforcement program. Penalties for unauthorized activities can be as high as $25,000 per day of violation, or imprisonment for up to one year, or both. But agencies seldom pursue civil or criminal actions against violators, even for egregious violations. A recent study by GAO on the number of civil or criminal actions taken by five Corps district offices found that only six civil and no criminal actions were pursued by the Corps over the three-year period from 1984 to 1986. The report found that the Corps devotes most of its time to processing permits, so monitoring and enforcement receive a low priority.[18] Thus, many unauthorized activities go undetected.

Typically, when the Corps discovers a permit violation or an illegal fill, it attempts to negotiate a settlement to remedy or rectify the problem. It prefers administrative to legal remedies because of the time and money saved. Until recently, it often issued an "after-the-fact" permit, which essentially gave the violator a permit to do what it had already done, and usually required mitigation of some of the damage done to the environment. But, in January 1989, EPA and the Corps signed a Memorandum of Agreement (MOA) establishing new policies and procedures for handling violations. The MOA states that EPA will be the lead agency for cases involving repeat violators and flagrant violations, for cases requested by EPA and for cases when the Corps recommends that EPA issue administrative penalties. For all other cases where wetlands are filled without a permit, or for violations of permit conditions, the Corps will be the lead enforcement agency. The MOA will also make it harder for the Corps to issue after-the-fact permits, which will not be issued until all administrative, legal, and/or corrective actions have been completed.

Following enactment of the Water Quality Act of 1987 (WQA), EPA and the Corps are no longer the toothless enforcers they once were. The act has given them a set of sharp, new teeth for enforcement. EPA, especially, is getting tough on those who play fast and loose with federal wetlands laws, and a number of unfortunate developers have already felt the agencies' bite. Now, EPA and the Corps can issue penalties more readily.

The WQA established two classes of penalties for illegally filling wetlands: class I for minor violations and class II for major violations. Class I penalties can range as high as $10,000 per violation, up to a maximum of $25,000; class II penalties can reach $10,000 per day, up to a maximum of $125,000. The amount of the penalty depends on a number of factors, such as the nature, circumstances, extent, and gravity of the violation; history of violations; knowledge of the permit requirements at the time of the violation; and economic savings or benefits as a result of the violation. As of mid-1989, EPA had already issued fines ranging from $15,000 to $125,000. And in January 1990, the U.S. Current Court of Appeals upheld the strictest penalty ever meted out to a developer who illegally filled wetlands: three years in jail and a $202,000 fine.

THE TAKINGS ISSUE: WETLANDS PROTECTION VERSUS PROPERTY RIGHTS

A number of developers who have been denied 404 permits from the Corps have sought relief from the courts, claiming that denial of a permit constitutes a taking because it restricts the profitable use of private property, even though the government does not actually take possession. With few exceptions, however, these developers have been unsuccessful. Most state and federal courts have held that denial of a permit or the imposition of permit requirements for development in wetlands does not constitute a taking as long as it does not eliminate all economic use of the property.

The Fifth Amendment of the U.S. Constitution prohibits the government from taking property without just compensation. Government may take property under its police power or eminent domain, but it must pay for it. Government may also, however, limit the use of property through regulation without actually taking possession. If such limitations go too far, they may effect a taking. Until recently, the remedy for takings was for the government to amend or withdraw the offending regulation or exercise its eminent domain powers. But in a landmark ruling in 1987, the U.S. Supreme Court held that if a land use regulation gives rise to a taking, even if only temporarily, the remedy is just compensation (*First English Evangelical Lutheran Church* v. *Los Angeles County*, 107 S.Ct. 2378 (1987)).

One of the first wetlands cases to address a takings claim was *Morris Land Improvement Co.* v. *Township of Parsippany-Troy Hills* (193 A.2d 232 (N.J. 1963)). In this New Jersey case, the court applied the so-called harm/benefit theory and held that a municipal ordinance that restrictively zoned a privately owned wetland amounted to a taking. Under this theory, land may be regulated to prevent a public harm, but not to confer a public benefit; otherwise it is a taking. For example, a land use regulation prohibiting industrial uses in residential districts prevents a harm because it prevents the industrial uses from harming a residential neighborhood. By contrast, a land use regulation requiring preservation of historic landmarks confers a public benefit.[19] The New Jersey court held that the purpose of the ordinance was to promote a public benefit (flood protection) rather than to prevent a public harm, and was therefore a taking of property without compensation.

The *Morris Land* decision, however, was later substantially qualified in New Jersey. Furthermore, in 1972, a Wisconsin Supreme Court decision, *Just* v. *Marinette County* (201 N.W. 2d 761 (Wis. 1972)), stood the harm/benefit rule on its head. The *Just* case centered upon a Wisconsin statute that required all counties to adopt a state-approved shoreland zoning ordinance. Marinette County's ordinance required a permit to fill wetlands that were located within a certain distance from navigable waters. The Just family, who owned property within the defined shoreland, claimed that the ordinance constituted a taking since it diminished the economic value of their land. The Wisconsin Supreme Court rejected Just's claim and found that the ordinance was a reasonable exercise of the police power to prevent harm to environmental resources. According to the court, the ordinance did not improve the public condition but only preserved the natural environment from being destroyed by unregulated activities. Therefore, no taking occurred.

In a landmark takings case, *Penn Central Transportation* v. *New York City* (438 U.S. 104 (1978)), the U.S. Supreme Court rejected the harm/benefit rule and adopted a "whole parcel" rule for taking cases. In *Penn Central*, the Court upheld a historic landmark designation of Grand Central Station in New York City and so prevented Penn Central from constructing a high-rise office building in the air space above the terminal.

Penn Central argued that although it could receive a reasonable return from the terminal alone, the landmark designation was a taking because it prevented the company from developing the air space above the station. In deciding whether or not a taking had occurred, the Supreme Court refused to separate the air rights from the terminal beneath and looked instead at the property as a whole. The Court accepted that the landmark designation imposed a severe restriction on Penn Central's use of its property, but refused to accept that it was therefore a taking. The Court noted that the designation benefited all citizens of New York City by "improving the quality of life of the city as a whole." And although the Court agreed that Penn Central was forced to shoulder the entire burden of the designation, the Court reasoned that, as a member of the community, Penn Central also received benefits from the designation.

The Supreme Court's refusal to split the property into discrete parcels and then decide if limitations on a particular parcel constitute a taking has significant implications for wetlands law. Many wetlands regulations restrict development in wetlands but allow development on the upland portion of a site. Under the "whole parcel" rule, such regulations will not be considered a taking if the property owner is left with some viable economic use for the remaining portion of the property.

Applying this rule in *Deltona Corporation* v. *U.S.* (657 F.2d 1184 (Ct.Cl. 1981)), the U.S. Court of Claims held that denial of a Section 404 permit for development on two of Deltona's five tracts of land in Florida was not a taking since Deltona could still develop the other three tracts. In *Deltona*, the court also applied a two-part takings test first adopted in *Agins* v. *City of Tiburon* (447 U.S. 255 (1980)) and later by the U.S. Supreme Court in *Keystone Bituminous Coal Association* v. *DeBenedictis* (108 S.Ct. 1232 (1987)). Under this test, a taking occurs if a regulation does not substantially advance a legitimate governmental interest or if a regulator leaves the landowner with no economically viable use of the property.

In 1964 and 1969, Deltona applied for and received permits to develop a residential community on two of five tracts of land it owned along the Gulf Coast of Florida. By the time Deltona applied for permits for the remaining three tracts in 1973, the Corps' jurisdiction had expanded considerably and the requirements for obtaining a permit were more stringent. As a result, the Corps granted Deltona a permit for only one of its three remaining tracts and denied permits for the other two. Deltona sued in the U.S. Court of Claims on the grounds that the Corps' permit denial constituted a taking. In striking down Deltona's claim, the claims court found that the Corps' regulations substantially advanced legitimate and important governmental interests and did not deprive Deltona of all economically viable use of its property. The court noted that for the two tracts on which a permit was denied, Deltona still had over 111 acres of uplands it could develop without a permit from the Corps.

The *Keystone* case confirmed the "whole parcel" rule. In that case, a Pennsylvania antisubsidence law placed restrictions on coal mining underneath houses and other buildings. A group of coal companies sued and claimed that the law "took" the coal that they would be required to leave in the ground. The Supreme Court, however, held that the law was enacted to restrict land uses that were "tantamount to [a] public nuisance." As in the *Penn Central* case, the *Keystone* Court looked at the entire parcel of land—all the coal companies' mines and not just the coal that would have to be left in the ground because of the regulation—and held that the law did not interfere with the coal companies' "investment-backed expectations."

By contrast to *Deltona*, another U.S. Court of Claims case has reached the opposite result. In *Florida Rock Industry, Inc.* v. *U.S.* (8 Cl.Ct. 160 (1985)), the court held that the Corps' denial of a permit to mine phosphate rock in wetlands in Florida constituted a taking because rock mining was the only economically viable use for the company's land. Florida Rock bought property in southern Florida in the early 1970s with the express purpose of mining phosphate rock. The court stated that "to leave the plaintiff with a commercially worthless piece of property in the name of preserving wetlands would be to charge the plaintiff with more than its fair share of this public cost" (*Florida Rock* at 177). The case is significant in that it was the first case to hold that denial of a 404 permit is a taking.

On appeal, the Federal Circuit Court upheld the claims court's ruling, but found that the court had erred in assessing the value of Florida Rock's property on the basis of current uses only (namely, mining) rather than considering other speculative uses. Although it remanded the case to the claims court, the appeals court stated that there is a "substantial possibility that a taking should be held to have occurred" (*Florida Rock Industry, Inc.* v. *U.S.* (791 F.2nd 905 (D.C. Cir. 1986)).

In a more recent case, in which a developer alleged that denial of a 404 permit was a taking, the U.S. Court of Claims strongly hinted that it felt a taking had occurred, but it held that further proceedings were necessary before it could reach a decision (*Loveladies Harbor, Inc.* v. *U.S.* (16 Cl. Ct. 381 (1988))). Loveladies Harbor applied for a permit to fill 12.5 acres of wetlands in its 51-acre site in Ocean County, New Jersey. The Corps found that one of 12.5 acres was actually uplands, so no permit was required, but it denied a permit for filling the remaining 11.5 acres.

After failing to overturn the Corps' denial in federal district and appellate courts, Loveladies brought an action in the U.S. Court of Claims. The court applied the same two-pronged takings test that was used in *Agins* and *Keystone*, but it discredited the "legitimate governmental interest" prong since "no court has ever found that a taking has occurred solely because a legitimate state interest was not substantially advanced" (*Loveladies* at 390).

The claims court determined that the 11.5 acres were worth about $4 million with a permit and only about $14,000 without and that the one acre of uplands was worth about $680,000. Given the severity of the economic impact, the court stated that it is "hard to imagine a takings claim more deserving of compensation" (*Loveladies* at 396).

The court, however, cited other decisions where courts held that a mere diminution of value is not sufficient grounds to find a taking and that plaintiffs must also be denied all economic use of their property before a taking can occur. The claims court stated that if, in future proceedings, Loveladies could show that the one acre of uplands cannot be developed, then the company will prevail at trial; if not, then the company's takings claim will be denied. Earlier, however, the court stated that the one acre of uplands is, in effect, "an island surrounded by a sea of wetlands," and that a taking will occur if all routes to that property are cut off. Thus, it appears that the court has set the stage for a finding that a taking has indeed occurred. (A trial was held in October 1989, but as of January 1, 1990, the claims court has still to reach a decision.)

Still in its infancy, wetlands takings law is constantly changing. Since the *Just* decision, most courts have given the federal government broad authority to regulate activities in wetlands. But the *First English* and *Nollan* (discussed below) decisions marked turning points in takings law in general, and, following *Loveladies*, as it applies to wetlands as well. It is unclear now just how far restrictions on wetlands development can go before they are considered a taking. If a taking is found in the *Florida Rock* and *Loveladies* cases, then the judicial pendulum will swing in favor of landholders, and in a relatively short time wetlands law will have come full circle: from *Morris Land* to *Just*, then to *Deltona* and *Riverside*, and, after *Loveladies*, back again to *Morris Land*.

In the meantime, when denied a permit from the Corps, developers are likely to be frustrated in their efforts to seek relief from the courts. When seeking permits, developers should be prepared to make concessions upfront and should expect to include some sort of mitigation. The onus rests with the applicants to prove to the regulatory agencies that not only have they done all they could to minimize the adverse environmental impacts of their project, but that their projects may actually enhance the environment. On the other hand, regulatory agencies must be careful about what conditions they place on a permit, and they must ensure that a reasonable connection exists between the mitigation requirements and a public purpose. Otherwise, they run the risk of a taking.

In March 1988, in a move that may have a profound effect on the 404 program, President Reagan issued Executive Order 12630, "Governmental Actions and Interference with Constitutionally Protected Property Rights" (53FR 8859 (March 18, 1988)). The Order directs federal agencies to "evaluate carefully the effect of their administrative, regulatory and legislative actions on constitutionally protected property" in order to "prevent unnecessary takings." Government regulatory programs, such as the 404 permit program, are included.

The Order stems from the landmark 1987 U.S. Supreme Court decision, *Nollan* v. *California Coastal Commission* (107 S.Ct. 3141 (1987)). Nollan, a landowner, had applied to the California Coastal Commission for a permit to expand his beachfront house. The commission issued the permit on condition that Nollan would grant an easement to allow public access to the beach. Nollan sued on the grounds that the requirement is a taking. The U.S. Supreme Court held that the essential "nexus" between the commission's purported goal of preserving beach access and the burden (easement) placed on Nollan's property was lacking, and that the exaction, therefore, amounted to an impermissible "taking."

The Order partly reflects the Reagan administration's concern that the government may be vulnerable to such takings claims, but it was also an attempt by the Reagan administration to limit government regulation of private property. In accordance with the Order, the attorney general developed guidelines that establish a uniform framework for federal agencies to use when evaluating the takings implications of their actions—the so-called "takings implication assessment." The assessment allows an agency to determine whether or not a proposed policy or action poses a risk of a taking and what the possible financial exposure of the proposal might be. Before placing any conditions on a Section 404 permit, for example, the Corps must perform an assessment and give special thought to the takings implications. Depending on how rigorously it is implemented, the Order could dampen federal enthusiasm to regulate wetlands. The Corps may be more reluctant to deny 404 permits and may instead grant more with mitigation.

The current method of processing permits on a case-by-case basis is inefficient. The rather clumsy administrative framework—where the Corps administers the program under EPA oversight—and the sheer volume of applications makes the process unwieldy for both the Corps and EPA alike. Indeed, nobody is particularly happy with the process that operates, as former Governor Thomas Kean of New Jersey put it, "a split personality."[20] Developers find the process unpredictable and often costly. Since the Corps and EPA have wide latitude to interpret the regulations, interpretations often vary across regions. A permit granted for a project in one area may be denied for a similar project under similar conditions in another. No hard and fast rules exist for developers to follow. They can never be sure if or when they will get a permit or how much mitigation will be required and how much it will cost. Environmentalists contend that the case-by-case process allows wetlands to be destroyed piece-by-piece, with individual projects slowly chipping away at a larger wetlands ecosystem and little thought given to cumulative impacts. A recent EPA review of over 90 Section 404 permits issued in Oregon and Washington found that most of the permits were for small fills of one acre or less.[21] These small fills add up, resulting in substantial cumulative wetlands losses.

EPA and the Corps have not been insensitive to such complaints. In 1989, EPA sponsored a study, conducted jointly by ULI–the Urban Land Institute and the Environmental Law Institute (ELI), of four regional planning efforts that endeavored to balance wetlands conservation with wetlands development. The four areas studied were Anchorage, Alaska; East Everglades, Florida; the Columbia River Estuary in Oregon; and the Chesapeake Bay in Maryland. The study found that, in general, regional wetlands planning improved the existing permit process by providing greater predictability for developers while offering conservationists greater assurance that certain wetlands would be protected. Based on their evaluation of the four regions, ULI and ELI identified a number of factors that were necessary for regional wetlands plans to be successful, such as strong leadership from a key public official to endow the planning effort or group with legitimacy, purpose, and authority; a method (or institution) and funding to implement the plan; availability of upland (nonwetland) areas that can accommodate growth; and participation of federal, state, and local agencies with jurisdiction over wetlands.[22]

In addition, in 1987, at the behest of EPA, The Conservation Foundation convened a National Wetlands Policy Forum to address some of the deficiencies in the current wetlands programs. The 20-member forum, chaired by Governor Kean, was composed of representatives from business, farming, and environmental groups, as well as from federal, state, and local governments. In November 1988, after more than a year of deliberations, the forum released an extensive list of recommendations for improving federal, state, and local wetlands programs. It recommended adopting both regulatory and nonregulatory approaches to improve the management and protection of U.S. wetlands. First and foremost, the forum recommended that the United States establish a national wetlands protection policy to achieve "no net loss" of wetlands over the short run and to increase the wetlands base over the long run. No net loss means that wetlands created will balance wetlands destroyed. The forum also recommended:

- expanding wetlands programs to cover all kinds of wetlands alterations, such as draining and excavation, and not just deposit of fill;
- implementing stronger mitigation requirements;
- expanding government wetlands acquisition and preservation programs;
- expanding monitoring and enforcement of wetlands regulations;
- developing incentives to protect wetlands; and
- delegating responsibility for all wetlands regulations to the states.

The forum's recommendations, which represent a consensus from the group, have not fallen on deaf ears. EPA, which had anxiously awaited the final report, has already begun to implement some of the recommendations. In January 1989, EPA released a Wetlands Action Plan that outlines the agency's strategy for implementing the recommendations. The agency has enthusiastically adopted the goal of no net loss and has pledged to toughen enforcement of wetlands protection laws and to require mitigation for all fills. Former EPA chief administrator, Lee Thomas, said that the action plan will send a "clear message that wetlands protection is a high priority." His successor, William Reilly, the former head of The Conservation Foundation, is likely to adopt additional forum recommendations.

ADVANCE IDENTIFICATION

A major flaw of the current wetlands regulatory system is that it is reactive rather than proactive; both EPA and the Corps spend most of their time responding to permit requests and very little time working to avoid conflicts in advance. Under section 230.80 of the EPA 404(b)(1) guidelines, however, EPA and the Corps may, based on their joint identification and evaluation of wetlands, designate certain wetlands as suitable or unsuitable for disposal of dredge or fill material in advance, before development plans or permit requests are imminent. Advance identification is proactive and it can help regulators and developers avoid unnecessary and costly conflicts. The designations do not grant or deny permits in advance; in general, applicants still must obtain an individual permit. But because the Corps and EPA have agreed in advance on where fills should and should not occur, conflicts between the two agencies should be fewer. Developers may still apply for a permit to fill wetlands designated as off-limits, but their chances of obtaining a permit are slim.

Advance identification can make the permit process more predictable, can reduce government regulation, and can provide a framework for reaching definite regulatory results (permits issued, permits conditioned, and permits denied) more efficiently than the case-by-case method. Moreover, advance identification usually bolsters public awareness and appreciation of wetlands values and allows the regulatory agencies to assess cumulative impacts of isolated fills on an entire watershed. As one EPA official noted, advance identification helps prevent wetlands from getting "nickel and dimed to death." Advance identification gives developers the predictability they desire while ensuring that sensitive areas (such as rare or pristine wetlands) will be protected and that the cumulative impacts of wetlands fills will be considered in the context of an entire ecosystem.

The disadvantages of advance identification include the difficulty of securing agreement among parties, the very resource-intensive nature of the process for the agencies involved, and the fact that it does not always work. For example, landowners whose land will fall under the off-limits category will likely object to such a designation, while conservationists may protest the classification of certain wetlands as expendable.

Between 30 and 40 advance identification processes have been proposed, completed, or are underway in such places as the Hackensack Meadowlands, New Jersey; Chincoteague Island, Virginia; York County, Maine; Pearl River Basin, Louisiana; East Everglades, Florida; Rainwater Basin, Nebraska; San Francisco Bay; along the Jordan River in Utah; and in the dunes region of Indiana.[23]

At the southern tip of Lake Michigan in northwest Indiana, the Grand Calumet River/Indiana Harbor Canal drains a heavily industrialized watershed. Much of the water in the area is polluted, and in 1985 EPA developed a master plan to improve the water quality of the river. Part of this plan involved wetlands preservation, so the agency decided to embark on an advance identification process. The area includes a region called

FIGURE 2.2
CLASSIFICATION OF WETLANDS IN ANCHORAGE, ALASKA

Key:
- Preservation
- Conservation
- Development
- Special Study

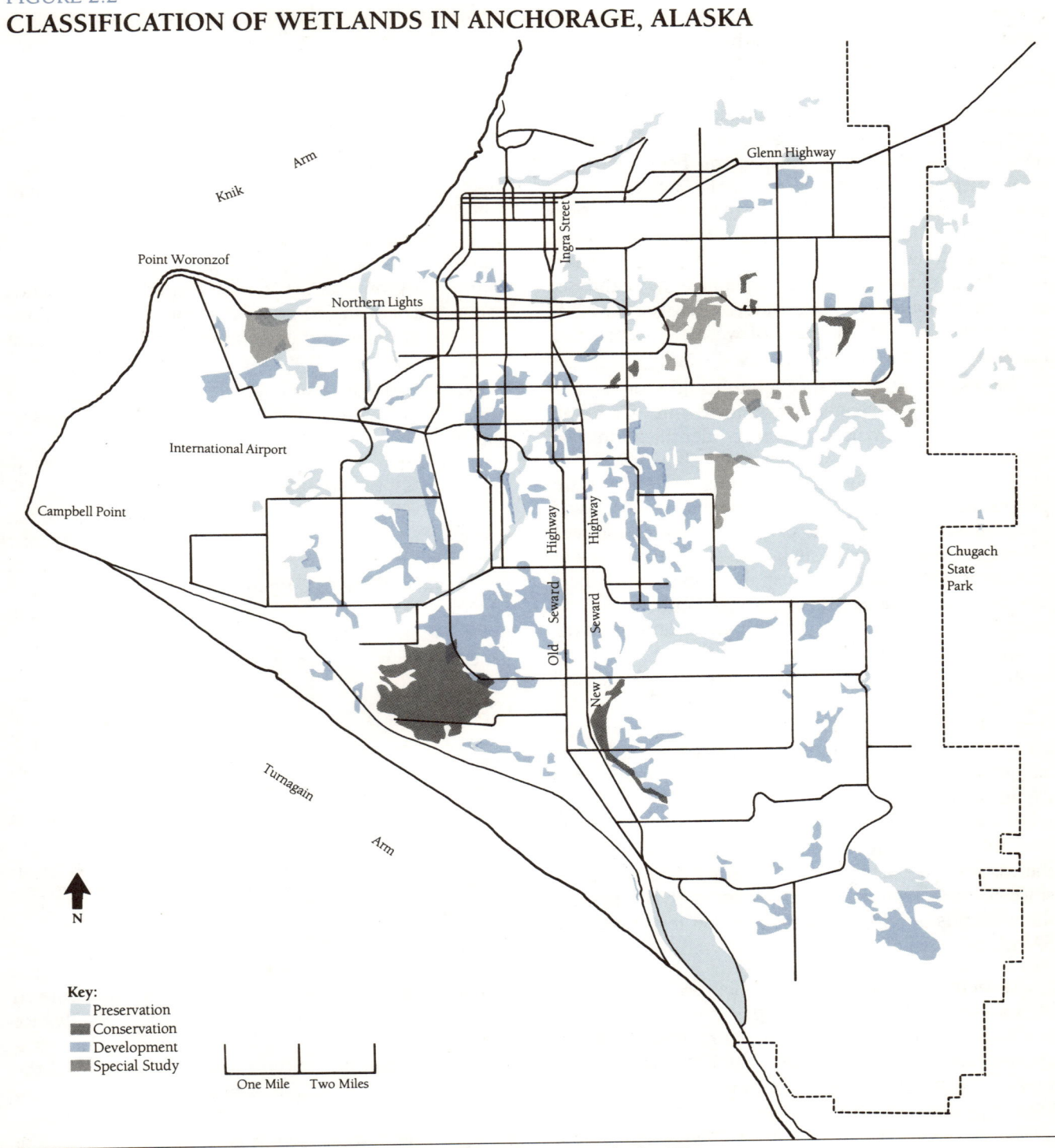

Source: Municipality of Anchorage Wetlands Study, Wetland Designations Anchorage Bowl, July 1987.

the "Indiana Dunes," which is characterized by a series of dunes and ponds along Lake Michigan. The ponds provide habitat for many endangered plants that are not found elsewhere in Indiana. The ponds were formed in the 1930s when the lake level and water table rose.

The advance identification process involved the Corps, FWS, the National Park Service, Indiana Department of Natural Resources, Northwestern Indiana Regional Planning Commission, the Lake Michigan Federation, and EPA. Wetlands were identified, mapped, and evaluated. Since many wetlands had already been destroyed or degraded, the agencies decided that any remaining wetlands that still perform beneficial functions, such as flood control, or that offer wildlife habitat would be designated off-limits to development. This designation did not include disturbed or excavated wetlands, wetlands that seldom contain standing water, or wetlands of less than five acres (unless they exhibited outstanding natural resource values). Altogether, 1,832 of approximately 9,190 acres of wetlands were designated as unsuitable for fill.

The results of the advance identification for the Calumet River region were incorporated into a regional master plan. Elsewhere, advance identification may evolve into a special area management plan (SAMP) and be incorporated into local planning and permitting programs. Under the 1980 amendment to the Coastal Zone Management Act, the Corps, in conjunction with federal resource agencies such as EPA and FWS, as well as state and local agencies, may develop a comprehensive plan, called a SAMP, to provide both natural resource protection and reasonable coastal-dependent development in a specific geographical area. The plan contains a comprehensive statement of policies regarding, and criteria to guide, land use. Ideally, a SAMP concludes with details of local, state, and federal approvals or restrictions of certain activities (fills) in defined wetlands. The program is designed for coastal areas, but the advance planning concept can equally be applied to inland waters. The SAMP process works best where wetlands face strong development pressure and where there is strong local interest in, and support for, the plan, since local or regional agencies ultimately implement the wetlands development/preservation features of the SAMP.

One of the more successful SAMPs was implemented in Anchorage, Alaska. Anchorage—a coastal community of 215,000—sits on a triangular peninsula flanked on two sides by the Knik and Turnagain Arms of the Cook Inlet and on the third side by the Chugach Mountains. Squeezed between the mountains on one side and the Knik and Turnagain Arms on the other, development is limited to the existing, vacant land within the Bowl and Eagle River, which is also where many wetlands occur. The Alaskan economy is heavily dependent on oil. Anchorage's economy, like that of Houston or Denver, is vulnerable to fluctuations in oil prices. In the late 1970s and early 1980s, an oil boom in Alaska triggered investment and development in Anchorage. This, in turn, intensified development pressures in the wetlands. At the same time, federal, state, and local groups sought to protect the area's critical wetlands.

These competing pressures spurred the city to develop a wetlands management plan. The main objective of the plan was to produce a strategy that would protect wetlands serving important ecological or hydrological functions and identify wetlands of limited ecological or hydrological values within which development could occur. Developed as a cooperative effort among local, state, and federal officials, as well as citizens' and environmental groups and property owners, the wetlands management plan is part of Anchorage's coastal zone management plan and its comprehensive development plan.

The plan classified wetlands into four categories (see Figure 2.2): preservation, in which no development, with few exceptions, can occur; conservation, an intermediate category in which some development may occur and some wetlands are preserved; developable, in which development can occur most readily; and special study, which includes wetlands that must be studied further before they can be classified into one or more of the previous three categories. The Corps' case-by-case Section 404 permit process governs development in the preservation and conservation areas; the Corps issued a general permit for the developable area.

Although some problems remain with the plan, it has been largely successful. In general, developers are pleased that they can develop within certain areas and that the permitting process is more predictable and expedient. For their part, environmentalists are pleased that the plan protects many critical wetlands. Since the plan was enacted in 1982, most development has occurred in wetlands classified as developable, some in conservation wetlands, and about 125 acres of preservation wetlands have been filled, primarily for public projects such as roads or ballfields.

Despite the apparent success in Anchorage and in a few other areas, advance identification has a mixed record. Compared to the case-by-case method, advance identification may save neither time, money, nor wetlands. For example, in the early 1980s, EPA and the Corps began a tortuous and controversial advance identification of wetlands in Chincoteague Island, Virginia, an eight-mile-long by one-mile-wide barrier island at the northern end of Virginia's coastline.

Once primarily a fishing village, the island now harbors a thriving tourist industry. Its year-round population of only about 500 people swells to well over 15,000 during the height of the summer, creating a burgeoning demand for additional motels, campgrounds, trailer parks, retail stores, and restaurants. With over 1,000 acres of wetlands on this slender island, development frequently involves filling wetlands. The goals of the advance identification process were to increase public awareness of wetlands values and of the 404 program, to reduce future disputes between EPA and the Corps, and to provide some predictability to developers by identifying those wetlands that were suitable or unsuitable for fill.

Many local residents, developers, and some public officials resented federal intrusion into their affairs and property owners vociferously objected to any federal designation that would exclude their land from profitable development. The uproar surrounding the advance identification process did create a widespread awareness of the 404 permit program, but it did not prevent a few residents from filling wetlands without a permit. EPA has taken enforcement action against these violators.

In contrast to the residents of Chincoteaque Island, residents in the Poconos in eastern Pennsylvania generally welcomed advance identification for three wetlands comprising a total of roughly 1,200 acres. The Poconos residents felt that unbridled development was ruining their natural resources. In fact, according to EPA, strong public support for wetlands protection has led to increased enforcement of illegal fills by EPA and the Corps.

Despite the checkered history of advance identification to date, the process does seem to promise that, in the long run, the regulation of wetlands will be more efficient, predictable, and consistent. Advance identification should also enable federal agencies to foresee, and therefore to minimize, the cumulative impact of wetlands losses. Developers too should benefit from a process that identifies in advance which areas are open and which are closed to development.

A CLEARER FUTURE?

Although the federal agencies, particularly EPA and the Corps, have fought pitched battles in the past over the purpose and scope of the 404 program, recent developments suggest that the agencies are entering a period of mutual cooperation. Their joint delineation manual, memorandum of agreement on a mitigation policy and on enforcement, and increased interest in advance planning all mark tell-tale signs of improved relations between EPA and the Corps. Recent court cases have helped clarify federal agency roles and jurisdictions, but they have also muddied the waters. The *Riverside Bayview* decision affirmed the Corps' jurisdiction over wetlands adjacent to waters of the United States, but the market entry theory in the *Attleboro Mall* decision unfortunately added confusion to the permitting process. Overall, however, the prospects look good for EPA and the Corps to continue working toward providing a more consistent, predictable approach to regulating wetlands and perhaps providing greater protection for the nation's wetlands.

Nonetheless, even with full cooperation and understanding among federal resource agencies, wetlands will still fall prey to a variety of unregulated activities, such as dredging or draining, that lie outside the Corps' jurisdiction. In fact, the Corps regulates only a relatively small share of wetlands-altering activities. At the risk of duplicating federal efforts, several states have stepped in to fill some of the loopholes in the CWA that allow unmitigated wetlands destruction to occur.

Notes

1. Address by Theodore Roosevelt at the laying of the cornerstone, House Office Building, Washington, D.C., April 14, 1906.
2. 53 FR 3120.
3. Malcolm Baldwin, "Wetlands: Fortifying Federal and Regional Cooperation," *Environment*, vol. 29, no. 7, 1987, p. 19.
4. The origins of the Clean Water Act may be traced to 1948, when Congress first provided money for construction of municipal wastewater treatment plants. But it was not until 1972, when Congress enacted the Federal Water Pollution Control Act Amendments (FWPCA), that the current framework of the CWA was established. The FWPCA was amended and renamed the Clean Water Act in 1977. The Clean Water Act was amended once again in 1987.
5. 33 U.S.C. Section 1344(a).
6. Emphasis added; 40 CFR Section 230.10(a).
7. Emphasis added; 40 CFR Section 230.10(c).
8. 33 U.S.C. Section 1344(c).
9. U.S. General Accounting Office, *Wetlands: The Corps of Engineers' Administration of the 404 Program* (Washington, D.C.: author, July 1988), p. 52.
10. CWA Section 101(a).
11. See 40 CFR Section 230.10(a)(3).
12. 40 CFR Section 230.3(q).
13. 40 CFR Section 230.10 (c)(1),(2),(3); (c).
14. 40 CFR Section 1508.20.
15. Lieutenant-Colonel Kit J. Valentine, "Corps of Engineers: Mitigation Responsibilities," in *Mitigation of Impacts and Losses: Proceedings of the National Wetland Symposium* (Berne, New York: Association of State Wetland Managers, May 1988), p. 35.
16. Mary C. Landin and Andrew C. Miller, "Beneficial Uses of Dredged Material: A Strategic Dimension of Water Resource Management," in *Transactions of the 53rd North American Wildlife and Natural Resources Conference* (Louisville, Kentucky: The Wildlife Management Institute, Washington, D.C., 1988), pp. 315–325.
17. Based on a personal conversation on July 20, 1988, with Douglas Ehorn, deputy chief, Water Quality Branch, Region V, U.S. Environmental Protection Agency.
18. U.S. General Accounting Office, *Wetlands*, p. 66.
19. Dan Mandelker, *Land Use Law* (2nd edition, Charlottesville, Virginia: The Michie Company, 1988), pp. 24–28.

20. See Thomas H. Kean, "Protecting Wetlands: An Action Agenda," *Environmental Forum*, vol. 6, no. 1, January/February 1989, p. 23.
21. Mary E. Kentula, et al., "Trends and Patterns in Section 404 Permitting in the Pacific Northwest," U.S. Environmental Protection Agency, Corvalis Research Laboratory; forthcoming in *Environmental Management*.
22. For further details, see the summaries of four case studies given in the Environmental Law Institute's *National Wetlands Newsletter*, vol. 11, no. 6, November–December 1989, pp. 9–16.
23. Information obtained during a personal conversation on February 28, 1989, with Tom Muir of the U.S. Environmental Protection Agency, Office of Wetlands Protection, Washington, D.C.

CHAPTER 3

STATE WETLANDS REGULATION

> *By different methods different men excel;*
> *But where is he who can do all things well?*
> — Charles Churchill

States have not stood by idly while the federal agencies develop programs to regulate the nation's wetlands. In line with the philosophy of "new federalism"—in which states play a more active, and the federal government a less active, role in policy making—many states have enacted their own legislation to fill the regulatory gaps in the federal 404 program and protect their remaining wetlands.

The resulting programs, no two of which are identical, vary from those that regulate a wide range of activities such as dredging or draining, to programs that provide tax incentives to protect wetlands permanently. Massachusetts, for instance, which was the first state to regulate wetlands and which now has one of the strictest wetlands protection laws in the nation, regulates not only dredge and fill in wetlands, but also removal and alteration. New Jersey, which has regulated coastal wetlands since 1970, but which only recently began regulating freshwater wetlands, regulates removal, disturbance, or dredging of soil, drainage or disturbance of the water level or water table, and discharge or fill. And one maverick state, Michigan, has even qualified to assume responsibility for administering the 404 program within its own borders.

In general, states regulate wetlands in two ways: indirectly, as part of broad regulatory programs such as the coastal zone management program or the water-quality certification provisions under Section 401 of the Clean Water Act; and directly, by enacting laws specifically to regulate activities in wetlands.

This chapter provides an overview of state wetlands laws that protect coastal and inland wetlands, describes how those programs are implemented, discusses state permit processes, and concludes with a more detailed look at six state programs covering a variety of activities in both coastal and inland wetlands. Those states are: Florida, New Jersey, Oregon, California, Michigan, and Massachusetts.

STATE ASSUMPTION OF THE 404 PROGRAM

In 1977, Congress directed EPA to turn part of the 404 program over to the states. Under Section 404(g) of the CWA, EPA may delegate administration of the 404 permit program for nonnavigable waters to individual states. The Corps would retain its authority over navigable waters within a state, and EPA would continue to oversee a state's program.

Before a state can take over the program it must first obtain EPA approval, which may take several years. In order for a state program to qualify, it must have several features in common with the federal program; a state must review permit applications in compliance with the 404(b)(1) guidelines, issue a public notice, respond to comments, and so on. State programs may be broader and more stringent than the federal program, but the "extra" coverage will not be subject to federal oversight or enforcement.

43

For a variety of reasons, states have been disinclined to take over the program. By the end of 1989, only one state—Michigan—had assumed administration of the program and only one other state—New Jersey—was seeking to do so. The 404 permit program's regulations are cumbersome, its requirements are stringent, and the incentives to assume its administration are few. For example, in a coastal state that administers the federal program, the Corps would retain jurisdiction over most of that state's coastal waters. Oregon seriously considered assuming responsibility of the 404 permit process in 1988, but concluded that since the Corps would retain jurisdiction over much of the state's wetlands, and since the state would have to spend an estimated $410,000 per year to run the program, it simply was not worth it.

Furthermore, despite Congress's directive, EPA has shown a reluctance to part with its oversight. Although —at least obstensibly—EPA joins with the Corps in supporting handing the permit program over to the states, the agency fears that state regulators could succumb to pressure from the building industry and compromise the program. "EPA is afraid that state [resource] agencies will become pimps for developers," observed Stan Geiger of Scientific Resources, Inc., a consulting firm in Lake Oswego, Oregon.[1] Such concerns may be exaggerated. Even after EPA delegates program administration to a state, the agency can rescind its delegation if a state fails to meet EPA standards. In addition, states may not issue a permit over EPA's objection.

In recent years, however, EPA has tried to accommodate itself to the spirit of new federalism. In 1987, the agency amended its regulations to make them less onerous. And in light of the 1988 recommendation by the National Wetlands Policy Forum to delegate the program to the states, EPA may well take further steps to induce more states to follow Michigan's lead. EPA believes that federal and state wetlands regulation should be a partnership; if Michigan's experience is any indication, EPA will not meddle in the administration of state 404 programs. In 1985, only 1 percent of Michigan's permit applications were reviewed by EPA, and in 1986, only 1.5 percent.[2]

State assumption of the federal 404 permit program would substantially reduce duplication and delays caused by overlapping state and federal wetlands regulations. But most states will not seek to assume responsibility for the program until the federal government improves the incentives to do so—for example, by providing states with financial and technical assistance. The National Wetlands Policy Forum recommended that Congress should enact legislation allowing EPA to delegate discrete portions of the 404 regulatory responsibilities to states and to provide states with the financial and technical assistance necessary to develop programs to carry out their delegated responsibilities.[3]

INDIRECT PROGRAMS

Most wetlands protection efforts on both state and federal levels started along the coasts and slowly worked their way inland. Coastal wetlands have always received more attention and therefore more protection than their inland cousins, even though freshwater wetlands comprise the bulk of all wetlands.

Nonetheless, until 1972, only a few states had laws to protect their coastal wetlands. California and Oregon had broad coastal zone management programs in place by then, while Massachusetts and Connecticut enacted laws specifically to protect coastal wetlands. But in 1972, the Coastal Zone Management Act (CZMA) provided the impetus for the rest of the coastal states to follow their lead. Now, all but a handful of coastal states administer federally approved coastal zone management programs.

Each state program is different, ranging from those in New York, Massachusetts, and Virginia, which simply packaged existing state programs into a form that fitted the federal criteria, to those in California and North Carolina, which enacted specific state laws to implement the federal program. In some states, local governments implement the coastal zone program, leaving the state government with a limited role. Usually the state will establish guidelines or standards that the local programs must meet. In other states, the state government maintains direct control over program administration. Many states, such as Washington, California, and North Carolina, require local governments to prepare coastal zone management plans, whereas in Florida and New York such plans are voluntary. In several northeastern states, no formal local role exists, and the state relies on direct control of the coastal zone management program. In South Carolina, local governments may prepare shorefront management plans, but the state maintains regulatory control.

Washington was the first state to have its coastal zone management program approved by the National Oceanic and Atmospheric Administration (1976). The heart of its program is its 1971 Shoreland Management Act. The act requires each local government to prepare a management plan and a permit system for future development within its coastal zone. Local permits are reviewed by the state. The act applies statewide, covering over 20,000 miles of shoreline, but is not limited to the coastal zone. It applies to all marine waters, lakes over 20 acres, streams with an annual flow of over 20 cubic

FIGURE 3.1
THE PROPORTION OF TIDAL TO NONTIDAL U.S. WETLANDS

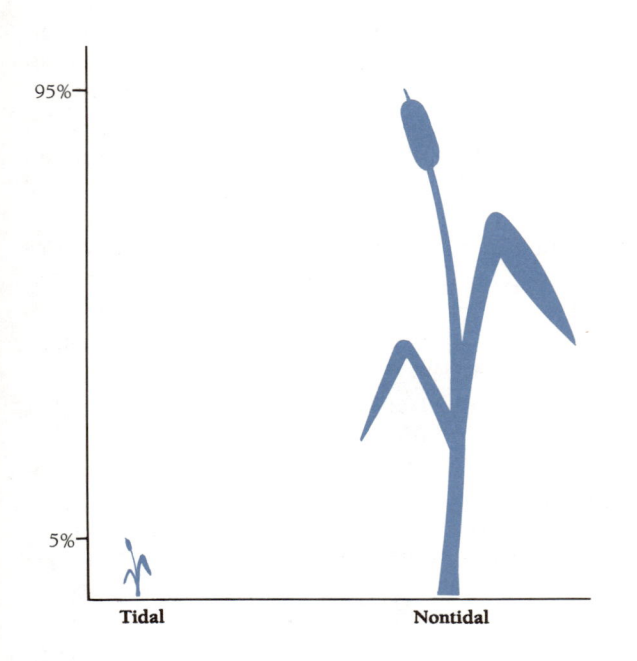

feet per second, and wetlands within 200 feet of these waters.

Ten years after Washington's program was approved, Virginia became the 29th of 35 eligible states to join the federal coastal zone management program. But unlike Washington, whose program is based primarily on one law, Virginia's program is based on a number of existing state laws, programs, and agencies. The state's Council on the Environment, which was established under its Environmental Quality Act, packaged the various state programs, such as the 1982 Wetlands Act that regulates activities in tidal wetlands only, into one comprehensive coastal program. The state can delegate management authority to local jurisdictions that adopt a model wetlands ordinance, which most of Virginia's local governments have done.

Several states enacted new legislation to create their coastal programs. For example, North Carolina's Coastal Area Management Act of 1974 became the basis for its coastal program. The act established the Coastal Resource Commission, which developed guidelines for local land use plans in coastal areas. Coastal counties must adopt plans that conform with state guidelines. Under the act, development is regulated in commission-designated "areas of environmental concern," which include coastal wetlands and estuarine waters.

State Water-Quality Certification

One of the main goals of the CWA is to reduce discharges of pollutants, particularly from industrial sources, into U.S. lakes, rivers, and streams. The 1972 amendments to the act established a permit program, known as the National Pollution Discharge Elimination System (NPDES), for point source discharges. The NPDES in turn established end-of-the-pipe standards for pollution discharges into U.S. waters. State water-quality standards, subject to EPA approval, are also used to control discharges. These standards can be used, indirectly, to prohibit development in wetlands.

Under Section 401 of the CWA, any federally permitted activity, including wetlands fills permitted under Section 404, must comply with state water-quality standards, which can be stricter than federal standards, just as state wetlands laws can be more stringent than federal laws. States can veto federally permitted projects that fail to meet their water-quality standards. Most states feel that Section 401 is not specifically designed to protect wetlands and that, therefore, it is not very effective in that role. In a few states, however, such as Iowa and Ohio, water-quality permits are a state's only means of protecting wetlands. In Ohio, for example, an antidegradation policy has been applied to water-quality certification in wetlands.[4]

Regional Programs

A handful of states have enacted programs that protect unique natural resources, including wetlands, in defined geographical areas. Maine, Florida, California, New Jersey, New York, and Massachusetts all have such programs. For example, in 1979, New Jersey enacted the Pinelands Protection Act to regulate development and protect natural resources, including wetlands, in a roughly 1 million-acre area in the south-central part of the state. Likewise, California's and Nevada's Lake Tahoe region also receives special regional protection. Florida's Critical Area Program applies to "areas of critical state concern," which may include an area containing natural, historical, or archaeological resources of regional or statewide significance. The program has been used to protect threatened resources in the Florida Keys and along the Suwannee River. Maine's Critical Areas Program was established in 1973 to identify and help ensure the protection of sites of unusual natural, scenic, or scientific value throughout the state, including exceptional plant or animal habitats, waterfalls, estuaries, beaches, and bogs. The program has three basic functions: to identify and document significant natural areas; to register them as "critical areas;" and to promote their voluntary conservation through cooperation with landowners. Since the program's inception, over 580 sites have been registered.

One of the more recent and well-known regional programs is Maryland's Critical Area Program, which was established in 1984 to protect the Chesapeake Bay.

Nearly 195 miles long and draining a 64,000-square-mile area that includes all of Washington, D.C., most of Maryland, Virginia, and Pennsylvania, and parts of New York, Delaware, and West Virginia, the Chesapeake Bay is the largest estuary in the United States. It had long been one of the most productive estuaries in the nation and was famed for its oyster and crab harvests. But pollution from a variety of sources both near and far began to take its toll on the Bay's resources. Oyster production declined, and a moratorium was placed on several species of commercially important fish, such as striped bass, that were contaminated with pollutants.

In response to continuing deterioration of the Bay, the Maryland General Assembly enacted the Critical Area Act in 1984. The act came on the heels of a five-year, $27 million EPA study that linked the Bay's decline to pollution from both direct sources, such as factories and power plants, and indirect sources, such as agriculture and residential development. It established a program to limit development within 1,000 feet—the critical area—of tidal waters or tidal wetlands in Maryland's portion of the Chesapeake Bay. All new development in the critical area must include a 100-foot buffer zone along tidal waters and wetlands and a 25-foot buffer for nontidal wetlands. Administered by local governments, which must adopt critical area plans that are consistent with criteria established by a statewide commission, the program aims to direct future growth away from the critical area and into areas that are already developed. The act directs local governments to map and classify their critical areas into three categories based on the density of existing land uses:

- Intense Development Areas—areas with a high concentration of residential, commercial, industrial, or institutional development. Development can occur most readily in this area, but developers must improve the quality of existing stormwater runoff and not allow additional development to impair water quality.
- Limited Development Areas—areas with moderate- or low-density residential development. Additional development can occur in this area, but it must not change the prevailing housing density of between one and four units per acre. To control stormwater runoff, development in this area must limit the amount of impervious surfaces within a site to 15 percent.
- Resource Conservation Areas—areas characterized by farms, open fields, wetlands, or forests. Development is limited to a maximum density of one unit per 20 acres. To ease the development restrictions on rural communities along the Bay that contain mostly resource conservation areas, up to 5 percent of such areas can be designated as either limited or intense development areas.

Although not designed specifically to protect wetlands, the critical area program, with its broad objectives to improve water quality and protect fish and wildlife habitat, contains a number of protective features for tidal and nontidal wetlands. For example, the program protects all tidal wetlands and nontidal wetlands "of importance to plant, fish, wildlife, and water quality." Development in the critical area must include buffers of natural vegetation along tidal waters and tidal and nontidal wetlands. Moreover, local governments must develop a wetlands mitigation plan for unavoidable, water-dependent impacts in nontidal wetlands.

Maryland's critical area program withstood its first legal challenge in 1988 when a developer argued that the Critical Area Act amounted to a taking of his property without compensation (*Meredith* v. *Talbot County*, No. CG 0510 (Md. Cir. Ct. July 14, 1988)). In that case, a developer submitted a proposal to subdivide 200 acres along a tributary of the Chesapeake Bay in Talbot County, Maryland. The county approved the proposal for all but 40 acres, which contained bald eagles. In order to meet a moratorium deadline for submission of development plans, the developer agreed not to build on the 40 acres, and based on this agreement, the county approved the project. Later, however, after the county rejected the developer's request for a waiver of the 40-acre restriction, the developer sued, arguing that he agreed to the restriction under duress. The court held that in making the voluntary agreement with the county, the developer was "not under such duress" that the court should void the agreement.

DIRECT PROGRAMS

Most states have enacted laws that apply specifically to wetlands (see table on pp. 48–49). The laws vary from those that authorize states to acquire and preserve wetlands to those that require permits for construction in wetlands. Typically, state wetlands laws authorize states to map wetlands and also to regulate certain activities in them. Maine was one of the first states to conduct a statewide inventory of its wetlands, although not every wetland was included in the inventory. Several other states, such as New York, Michigan, and Wisconsin, are in the process of mapping their wetlands; Connecticut has already mapped all of its inland wetlands. About half of the states have some sort of program authorizing them to purchase and preserve wetlands. For example,

Illinois's Natural Heritage Program established a trust fund that matches, dollar-for-dollar, private contributions to protect natural areas, including wetlands. Some states provide tax advantages, including reductions in property tax, income tax, gift or inheritance tax, or capital gains tax, to individuals who protect wetlands. A few other states offer property tax abatement for landowners who retain wetlands in their natural condition.[5]

Most of the 30 coastal states regulate development in coastal wetlands, but only 14 regulate freshwater wetlands as well. Some states, such as Louisiana and Mississippi, recently enacted coastal wetlands permit programs, but neither one has an inland wetlands protection act. Only one noncoastal state, North Dakota, has enacted a law requiring permits for certain activities in wetlands. In 1987, it adopted a no-net-loss policy that requires acre-for-acre replacement of wetlands drained after its wetlands law was enacted.

Some states modeled their programs after the federal 404 program and incorporated the same definitions, exemptions, and permit requirements as those employed by the Corps and EPA; other states adopted programs that extend far beyond the regulatory reach of the federal program and regulate more than just the deposit of dredge or fill. A few state wetlands regulations cover activities not only in a wetland itself, but also in a buffer strip around the wetland. Several states, such as Florida and Massachusetts, protect their coastal and inland wetlands under a single law, while others, such as New Jersey, have separate laws for each.

No two state wetlands programs are administered in the same fashion. In New England, where home rule reigns, states typically allow their wetlands program to be delegated to local governments. In other regions, control is retained at the state level. In Connecticut, municipalities once were only encouraged to regulate wetlands; now, following amendments to Connecticut's Inland Wetland Law in 1987, they are required to do so. In Maine, the state may delegate administration to municipalities, while Massachusetts's program is administered by over 300 conservation commissions but overseen by the state.

State definitions of wetlands vary, and so, therefore, do their jurisdictions. Although some states have adopted EPA's definition of wetlands, the majority have developed their own. Most states rely on vegetation or soils to establish wetlands boundaries, in contrast to EPA and the Corps, which use hydrology, plants, and soils. Connecticut, for example, relies on different soil types to separate wetlands from uplands. Massachusetts defines a wetland as an area where at least 50 percent of the vegetation is comprised of characteristic wetlands plants. North Dakota defines a wetland as a "natural depressional area that is capable of holding shallow, temporary, intermittent, or permanent water."

Residential development has all but engulfed this estuarine wetland in Long Island, New York. What is left remains scarred by mosquito control ditches. Most coastal wetlands filling in New York occurred before the state enacted its Tidal Wetlands Act in 1972.

States have concocted a variety of measures to delineate where the wetland ends and the upland begins, which is a bit like drawing a line separating the Gulf of Mexico from the Atlantic Ocean. Wetlands seldom establish clear, well-defined edges. Instead, the borders of wetlands can typically be characterized as a transition zone, a continuum between wetland and upland where the plant cover gradually shifts from predominantly wetlands plants to predominantly upland. Many plant species can tolerate a broad range of soil moisture conditions and so can grow in both wetlands and uplands. Black spruce grows in forested wetlands as well as in upland, spruce/birch forests. In other words, one cannot rely solely on plants to differentiate between uplands and wetlands.

The inconsistency of definitions and overlapping of state and federal jurisdictions often cause problems for members of the regulated community, particularly those members who have projects in different states. What is regulated in one state may not be regulated in another. And within a single state, local wetlands ordinances can vary from town to town. Although this lack of uniformity may complicate things for developers, it has certainly been a boon to wetlands scientists, who are often asked to delineate wetlands to avoid or settle disputes between developers and regulators.

State Permitting

The permit process for most state wetlands programs is very similar to the federal 404 program. Like the 404

STATE WETLANDS PROGRAMS AT A GLANCE

Alabama Permits required under its Coastal Area Management Act for activities (dredging, dumping, etc.) that alter tidal movement or damage flora and fauna.

Alaska Regulates activities in its coastal zone under the state Coastal Management Act. Federal lands exempt.

Arizona No specific wetlands protection program.

Arkansas No specific wetlands protection program.

California Statewide Coastal Commission regulates all development activities in the coastal zone, except around San Francisco Bay, which is regulated by BCDC. *See pp. 55–59.*

Colorado No specific wetlands protection program.

Connecticut Permit required for just about any alteration of coastal or inland wetlands, including dredging, removal, fill, and construction. All wetlands mapped. Program administered by local governments. Most agricultural activities exempt.

Delaware Regulates activities in coastal wetlands, including dredging, filling, bulkheading, etc., under its Wetlands Act. Essentially forbids construction of private, nonwater-dependent projects in tidal wetlands. Delaware is developing a freshwater wetlands protection law.

Florida Regulates activities (dredge and fill) in both freshwater and coastal wetlands. *See pp. 50–52.*

Georgia Regulates activities (dredge, fill, and drain) in salt marshes under its Coastal Marshlands Protection Act.

Hawaii Regulates development (dredging, removal, grading, construction, etc.) in the coastal zone under its Coastal Zone Management Act. Permits required from county coastal management authorities. Establishes shoreline setbacks of between 20 and 40 feet for new construction.

Idaho No specific wetlands protection program.

Indiana No specific wetlands protection program.

Illinois No specific wetlands protection program. Regulates some activities in floodways under its Rivers, Lakes, and Streams Act of 1911.

Iowa No specific wetlands protection program. Has active wetlands acquisition program by which state purchases and restores wetlands, primarily prairie potholes. Wetlands protection bill introduced to legislature.

Kansas No specific wetlands protection program.

Kentucky No specific wetlands protection program.

Louisiana Under its State and Local Coastal Resources Management Act, state and/or local permits required for activities (dredge and fill) in coastal wetlands.

Maine Permit required for activities that affect "protected natural resources," including coastal and inland wetlands. Freshwater wetlands under 10 acres exempt. Also, local governments establish setbacks for developments along freshwater and coastal wetlands.

Maryland Two programs: one regulating dredge and fill of tidal wetlands, the other regulating a wide variety of activities (such as removal, alteration, destruction of plants, grading) in freshwater wetlands. Both programs generally exempt agriculture and forestry. Established a no net loss policy for nontidal wetlands.

Massachusetts Wetlands program administered by local conservation commissions. Regulates activities (removal, fill, dredge, and alteration) in both freshwater and coastal wetlands. *See pp. 64–67.*

Michigan The only state to "assume" the federal 404 program. Regulates development in wetlands under a variety of programs. Generally exempts agriculture and recreation. *See pp. 62–64.*

Minnesota Permits required for any work in wetlands, under its Protected Waters and Wetlands Permit Program. Regulates lakes, ponds, cattail marshes, and open water marshes over 10 acres in rural areas and over 2.5 acres in cities. Certain uses prohibited outright, such as filling wetlands for a parking lot. Generally exempts agricultural drainage.

Mississippi Regulates dredging, dumping, filling, destruction of flora, and construction in coastal wetlands. Many activities exempt.

Missouri No specific wetlands protection program.

Montana No specific wetlands protection program.

Nebraska No specific wetlands protection program.

Nevada No specific wetlands protection program.

New Hampshire Permit required for any alteration of coastal or freshwater wetlands. Regulations stricter for coastal wetlands.

New Jersey Regulates development in both freshwater and coastal wetlands. Freshwater wetlands program similar, but broader, than the 404 program. *See pp. 52–55.*

New Mexico No specific wetlands protection program.

program, most states that require a permit for filling or draining wetlands evaluate proposals on a case-by-case basis and approve most permit applications upon the condition of some form of mitigation. In addition, states generally spend most of their limited resources reviewing permit applications and therefore skimp on monitoring and enforcement. Indeed, very few projects are actually monitored. For example, Florida has only one full-time staff person who reviews monitoring reports of mitigation sites.

New York Under a variety of laws, the state regulates development in freshwater and tidal wetlands. Generally, the freshwater wetlands program applies to wetlands of 12.4 acres and larger, all of which the state has mapped. Freshwater wetlands of less than 12.4 acres are covered if of "unusual local importance." Local governments can assume administration of freshwater program, but few have. Agricultural exemptions.

North Carolina Under its dredge and fill act, a permit is required to fill or excavate tidal wetlands. In addition, under its Coastal Area Management Act, the state also regulates development in areas of environmental concern, which include wetlands, estuaries, and floodplains within the coastal zone.

North Dakota Wetlands program focuses on agricultural drainage. Permits required to drain a wetland within a watershed of 80 acres or more. Requires replacement of drained wetland on a one-for-one basis.

Ohio No specific wetlands protection program. Ohio's "antidegradation policy" under its water-quality standards requires mitigation for wetlands alterations permitted by the Corps.

Oklahoma No specific wetlands protection program.

Oregon Under its Fill and Removal Act, a permit is required to fill or remove any material from "waters of the state," which include inland and coastal wetlands. In addition, local governments incorporate statewide planning goals, which include wetlands protection. *See pp. 59–62.*

Pennsylvania Under its Dam Safety and Encroachments Act, the state regulates encroachment on bodies of water, which includes draining, filling, or enlarging wetlands. Regulations are more stringent for "important" wetlands. Exemptions for cutting vegetation.

Rhode Island The state regulates development in both coastal and freshwater wetlands. Coastal wetlands are more stringently regulated than freshwater wetlands. Coastal program establishes six wetlands categories and identifies permitted uses in each. Freshwater program exempts small freshwater wetlands (i.e., swamps under three acres or marshes under one acre).

South Carolina Under its Coastal Management Act, the state regulates activities (dredge, fill, drain, etc.) in "critical areas," which include coastal waters and tidelands. Freshwater wetlands unregulated by the state.

South Dakota No specific wetlands protection program.

Tennessee No specific wetlands act, but the state regulates any alteration to "waters of the state," including wetlands, under its Water Quality Control Act. No development allowed in outstanding wetlands. Most agricultural activities exempt.

Texas No specific wetlands protection program.

Utah No specific wetlands protection program.

Vermont The state Water Resources Board designates "wetlands of state significance" and, instead of requiring permits, the board establishes allowable uses in those wetlands. The board is authorized to regulate activities that threaten state-protected values such as flood control, water quality, wildlife habitat, and aesthetics.

Virginia Regulates activities in coastal wetlands under the Wetlands Act. Permits issued either by state or by local governments that adopt the state's Wetland Zoning Ordinance. The act provides standards and policies for evaluating wetlands development proposals.

Washington Under its Shoreline Management Act, the state regulates development in waters of the state, including wetlands. Although not the main focus of the act, the state has jurisdiction over wetlands associated with tidal areas and over large streams and lakes. Permits issued by local governments, with final approval required by the state. Only local approval needed for very small projects (e.g., projects with a market value under $2,500).

West Virginia No specific wetlands protection program.

Wisconsin Wisconsin's Shoreland Management Program requires each county to adopt state-approved zoning ordinance for shorelands, defined as 1,000 feet from lake or pond and 300 feet from river or stream. The ordinance includes a shoreland–wetlands zoning district, which permits certain activities such as recreation and forestry, and prohibits all others, such as dredge and fill.

Wyoming No specific wetlands protection program.

Sources: Personal conversations with state wetlands program managers— AK, AR, AZ, CA, CO, CT, DE, FL, ID, IL, IN, IA, KS, KY, ME, MD, MA, MI, MN, MO, MT, ND, NE, NJ, NM, NV, NY, OH, OK, OR, PA, RI, SD, TN, TX, UT, VA, WA, WV, WI, WY.
William, L. Want, *Law of Wetland Regulation* (New York, New York: Clark Boardman Company, Ltd., 1989)— AL, GA, HI, LA, MS, NH, NC, SC, VT.
Note: Those states without a specific wetlands protection program generally rely on federal regulation under Section 404 and state water-quality certification under Section 401 of the Clean Water Act.

Developers must obtain both state and federal approval for projects in wetlands, and sometimes the process can be annoyingly duplicative. In general, the Corps will not issue a 404 permit unless the proposed project complies with state laws, including state water-quality certifications and state wetlands laws. Some states, such as Florida, Georgia, Mississippi, and Pennsylvania, conduct joint permit processing with the Corps, and this helps reduce processing time. Other states are headed in the same direction. In North Carolina, if the state issues a permit for dredging and filling in a wetland, the Corps will usually follow suit.

State Mitigation Policies

Only a few states have established formal mitigation policies. Florida usually does not allow off-site mitigation, but it does allow "preconstruction mitigation," that is, mitigation banking. In addition, every mitigation project in Florida above one-tenth of an acre must be put into a perpetual conservation easement to ensure that the mitigated site itself will not be the site of future development. A few state wetlands laws specify the type of plants and the amount of plant cover required at mitigation sites. For instance, in Massachusetts, at least 75 percent of the surface area of the replacement area must be established with native plants.

Some state mitigation policies require a certain minimum ratio of wetlands created to wetlands lost. For example, in New Jersey, the ratio is 2:1, although the state allows mitigation at less than 2:1 (but not less than 1:1) under certain circumstances. In California, the ratio is at least 1:1 and can be considerably higher. In South Carolina, the ratio can be as high as 3:1. Florida usually requires at least 2:1 mitigation, but the ratio varies with each project depending on the likelihood of success, geographical location, and whether wetlands will be created or enhanced. Connecticut's coastal wetlands law is so strict that it does not need a mitigation policy; since 1969, only about five acres of coastal wetlands have been filled.

To avoid costly delays and design modifications, applicants should meet with federal and state agencies to discuss their proposed plans early in the planning stage. It is not unusual for a developer to redesign a project at considerable expense to satisfy a state's requirements to protect wetlands, only to find out later that the project must be modified again to meet the requirements of the Corps or EPA.

SIX EXAMPLES OF STATE WETLANDS PROGRAMS

Of those states administering comprehensive wetlands programs covering activities in both coastal and inland wetlands, a handful stand out as having the toughest, most innovative programs in the country. For example, one state generally allows no more than 5,000 square feet of fill to be placed in a wetland, another has developed a formula to calculate the amount of mitigation required for a particular wetland, and a third classifies wetlands into three categories and sets different development standards for each. This section provides a brief overview of six of these states, whose wetland programs give some indication of where other states may be heading.

Florida:
All in the Coastal Zone

Florida is one of only two states—the other is Delaware—located entirely within the coastal zone as defined by the CZMA.[6] Over 90 percent of the state is within 75 miles of the coast, and its highest point is only 345 feet above sea level; no wonder the state is covered with wetlands. Historically, the state's attitude toward its wetlands was inhospitable. Its philosophy, in short, was dike, ditch, and drain. Florida has a long history of diking, draining, and filling its wetlands for agricultural and urban development. Many of its wetlands and rivers were permanently altered by canals and dikes that crisscross the state. But the Warren S. Henderson Wetlands Protection Act of 1984 turned things around. The act states that "while state policy permitting the uncontrolled development of wetlands may have been appropriate in the past, the continued elimination or disturbance of wetlands in an uncontrolled manner will cause extensive damage to the economic and recreational values which Florida's remaining wetlands provide."[7]

The act directs the Department of Environmental Regulation (DER) to regulate dredging and filling in surface waters of the state, defined as rivers, streams, bays, bayous, sounds, estuaries, lagoons and their natural tributaries, and natural lakes—except those owned entirely by one person or those that become dry each year and are without standing water. The act exempts numerous activities such as maintenance dredging of existing canals, installation and maintenance of boat ramps on artificial bodies of water, construction of seawalls and private docks on artificially created waters, installation of transmission lines, and certain agricultural activities.

DER is not the only state agency regulating activities in wetlands. Florida's Water Management Districts (WMDs) manage the state's surface waters for drinking water and to control floods. DER's jurisdiction extends to excavated water bodies that are connected to waters of the state, but it excludes isolated wetlands. Often, what DER does not cover, a WMD does. Thus, where isolated wetlands of half an acre or greater play a part in water supply or flood control, they fall under the WMD's jurisdiction. In addition, DER may delegate the program administration to the WMDs, as it did in the St. Johns District. And counties may enact wetlands regulations that are broader or stricter than the state's; Hillsboro County, for instance, regulates activities in isolated wetlands.

Permitting

Florida has two kinds of wetlands permits: standard permits for fills over 10,000 cubic yards, and short form permits for fills under 10,000 cubic yards. Both permit applications are filed with DER and both are subject to the same review standards. Upon receipt of an application, DER has 30 days to determine if the application is complete. According to one developer, DER invariably asks for additional information, and on major projects it is not unusual for developers to receive a second request for additional information. According to developers, these demands can add up to 90 days to the processing time. "We have prepared dozens of applications for developers and have never scored an A+ with DER, although we have with the Corps," noted one consultant. "DER always comes back with more questions."[8]

Within 24 hours of receiving a completed application, DER forwards a copy to the Corps and to the county and municipality (and to the Department of Natural Resources (DNR) if state lands are involved) and issues a public notice, usually within 30 days. The Corps may hold a public hearing, and DER may hold an administrative hearing. Generally, DER has 90 days to reach a decision or the application will be automatically approved. Permits are usually issued in 60 to 90 days, but the process may take longer (a year or more) for large or controversial projects. State and federal permit applications are processed simultaneously, but the Corps will not issue a permit if the state permit is denied.

Since 1984, DER has issued about 2,000 permits per year, and annual wetlands losses have ranged from 800 to 1,100 acres. But substantial gains in wetlands acreage have also been achieved. From October 1, 1987, to September 30, 1988, DER issued 2,013 permits that allowed about 933 acres of wetlands to be filled or dredged. Most permits covered alterations of less than five acres. Over 28,000 acres were created, however, during the same period. The tremendous gain in new wetlands resulted from a couple of very large government projects. A permit issued to the Florida Department of Transportation accounted for the largest single loss of wetlands, and a permit issued to a WMD was responsible for the largest single gain. The WMD will recreate over 26,800 acres of wetlands from wetlands that were once diked and converted to agriculture.[9]

In order to obtain a permit, a proposed project must pass two statutory tests: it must not violate state water-quality standards; and it must not be contrary to the public interest. The state public interest test is somewhat like the Corps' and includes an evaluation of the likely impacts of the project on public health and safety, fish and wildlife, navigation, and historical and archaeological resources.

If a project fails to pass the two tests, DER must deny the permit. But if the applicant is willing, and most applicants are, DER will explore a number of project modifications that would reduce or eliminate adverse environmental impacts and allow the project to receive a permit. If, after considering all practical modifications, the project still cannot meet DER's requirements, DER will then entertain proposals to offset the remaining adverse impacts.

Mitigation

DER will consider mitigation only when a project has failed one of the two statutory tests. In other words, mitigation projects will not be considered upfront. DER cannot actually require mitigation, but it does not have to issue a permit either. The goal of mitigation proposals is to offset expected adverse impacts to the point at which projects can qualify for a permit. Each mitigation proposal must include:

- A description of the mitigation area, including a description of the wetland to be created, enhanced, or protected, the type and functions of the wetland, species present, type of vegetation to be planted, plant density, source of plants, soils, and hydrologic regime.
- A planting plan that includes a species list, the proposed plant elevations and plant density for each species, and the source of plants (for example, from a nursery or from adjacent marshes).
- A monitoring plan that includes a timetable for monitoring and reporting, methods of statistical analysis, and criteria for success.
- A description of construction methods, including the equipment to be used, access methods and locations, and site preparation.
- A mitigation cost estimate. If the estimated cost exceeds $25,000, the proposal must include a description of the costs of earthmoving, planting, consultant fees, and monitoring.
- Contingency plans, in case the mitigation is unsuccessful.

In evaluating mitigation proposals, DER will consider the applicant's previous track record with developments in wetlands, the likelihood of success of the mitigation, and whether or not the applicant has sufficient legal interest in the mitigation site and the resources to guarantee that the mitigation will work. In certain circumstances, mitigation proposals will be unable to offset a project's adverse impacts sufficiently to warrant granting a permit; examples might include the presence of endangered species or the probability that a particular wetlands type will not be successfully created.

Permits always specify the mitigation activities that the applicant must perform, the criteria used for determining the success of the mitigation, and the monitoring requirements. DER's very specific criteria for success depend on the type of wetlands being created or enhanced. The criteria usually specify the minimum amount of plant cover required and the maximum amount of exotic plants or weeds allowed on site. For example, a manmade herbaceous wetland will be considered successful when plant cover reaches at least 80 percent of that found in an undisturbed, natural herbaceous wetland that serves as a control; aggressive, weedy plants such as cattail and primrose willow occupy 10 percent or less of the total cover; and the created wetland contains 75 percent of the plant diversity of the control wetland. Similarly, forested wetlands must have an average of at least 400 trees per acre growing above the herbaceous stratum, tree cover shall exceed 33 percent of the total area, and in no single area shall tree cover be less than 20 percent.

DER recognizes essentially two types of mitigation—creation and enhancement—and has established different requirements for each. For created wetlands, the ratio of wetlands created to wetlands destroyed is usually at least 1:1 but depends on the quality and rarity of the wetland, the likelihood of success, and how long it will take before the wetland is established. For example, the ratio for cordgrass marshes or red mangrove swamps could be 1:1, since these wetlands have been created with a reasonable degree of success. But for hardwood swamps or everglade wetlands systems the ratio would be higher to reflect the greater level of uncertainty.

No fixed ratio exists for enhanced wetlands, but DER generally starts at 4:1 and has required as much as 20:1. Enhancement is defined as improving an existing altered or affected wetland by such means as removing exotic or nuisance plants and successfully planting indigenous wetlands species. The mitigation ratio depends on the following factors: the degree to which the enhanced wetland has been stressed; type of stress; the cause of stress; and the likelihood of success of the proposed mitigation. DER believes that, given the limited success of many wetlands creation techniques, the state should encourage enhancement over creation. But, like New Jersey, DER generally requires a greater mitigation ratio for enhancement because it is much cheaper than creation.

Generally, off-site mitigation is acceptable in Florida only if on-site mitigation is not feasible and only if the off-site mitigation is performed in the same water body or within the same drainage basin as the fill site. "In-kind" mitigation is preferred but not required.

DER will also consider as mitigation land conveyances and conservation easements or other such deed restrictions, especially if such actions would preclude development in wetlands that are not protected by DER. DER may also require that the mitigation area be placed in a conservation easement if the site is outside DER's jurisdiction and if DER is not reasonably confident that the applicant will protect the area from development once the mitigation requirement is fulfilled.

According to a DER official, a major problem with the program is that developers often do not do the mitigation work that they said they would. In Florida, it costs about $45,000 an acre to create a wetland, including planning, grading, planting, and monitoring. "Developers can save a lot of money by skipping the mitigation, and some of them do," the DER official complained. In response, DER recently issued a mitigation rule that states, among other things, that for all mitigation estimated to cost over $25,000, developers must put up a performance bond to guarantee that the mitigation will actually occur. In addition, applicants may have to provide evidence that they have the financial resources necessary to correct whatever flaws may arise in the mitigation. Most permits require that the mitigation be performed concurrently with development, but developers with a bad track record will have to do the mitigation work upfront.

Enforcement

Florida devotes very few resources to enforcement. In the state's southwest district, 13 people issue permits but only one checks for compliance—and that person only looks at reports and does not perform field inspections. Field work is undertaken by the five district offices, but only one district (the southwest district) has a full-time permit compliance person. In 1988, the southwest district issued about 900 permits; only about 18 percent of these were monitored for compliance.

Comments

Increasingly, developers are doing such a good job of avoiding wetlands areas that Florida worries about losing too much of its valuable upland, such as mahogany hammocks. One of the problems with Florida's wetlands program, according to a developer, is that DER often becomes preoccupied with preserving even very small wetlands at all costs, while allowing a large, equally valuable upland forest to be destroyed. Upland habitat mitigation is on the horizon. According to DER: "The state needs to address loss of upland habitat before it is too late. If it does not, it will be playing catch-up like it did with wetlands."

New Jersey:
Innovative Approach to Wetlands Regulation

New Jersey has two wetlands programs: one for coastal (or, more accurately, tidal) and one for freshwater wetlands. The coastal program began in 1970 following enactment of the Wetlands Act, and was expanded in 1973 with enactment of the Coastal Area Facility Review Act (CAFRA). The freshwater wetlands program began only a few years ago with the Freshwater Wetlands Protection Act (FWPA) of 1987. New Jersey's coastal and freshwater wetlands programs, administered jointly, are much broader than either the federal Section 404 program or most state programs.

In two regions of New Jersey, the Hackensack Meadowlands and the Pinelands, wetlands are exempt from both the freshwater and coastal wetlands programs. Both the Meadowlands—a vast, marshy area in northeastern New Jersey—and the Pinelands—a large, ecologically unique and relatively undeveloped area in south-central New Jersey—are governed by special commissions that develop comprehensive plans and regulate growth in their respective jurisdictions.

The (coastal) Wetlands Act of 1970 applies to dredging, filling, draining, removal, and excavation of material from coastal wetlands, which are defined as "any bank, marsh, swamp, meadow, flat or other low-lying land subject to tidal action. . . ." FWPA covers many of the same activities as the Wetlands Act, but it also applies to any destruction of plant life, including tree cutting, that would alter the character of a freshwater wetland. CAFRA applies to a broad range of activities in the coastal area—which runs from the Raritan Bay down to Cape May, around the Delaware Bay and up along the Delaware River—such as construction of power plants or waste-treatment facilities, as well as construction of 25 or more housing units.

FWPA is much longer and more detailed than either of the coastal acts. It reflects former Governor Tom Kean's strong interest in wetlands protection and the extensive public debate that led to the bill's enactment. Kean, who chaired The Conservation Foundation's National Wetlands Policy Forum, pulled out all the stops to get the freshwater wetlands law passed. In June 1987, he imposed a temporary building moratorium in freshwater wetlands until an uncooperative legislature enacted a comprehensive wetlands bill. By the end of the month, the legislature had conceded, and the moratorium was lifted when Kean signed the bill on July 1, 1987.

Kean used similar tactics to try to coerce a recalcitrant legislature into creating a coastal commission that, like California's, would have broad authority to regulate development along the coast. In October 1988, Kean issued a "certificate of imminent peril" that put emergency regulations into effect for 60 days and halted construction of housing in coastal areas. Digging all the way back into New Jersey's legislative history to the Waterfront Development Act of 1914, the governor found the authority to regulate coastal development within 50 feet of the normal high-water mark. In November 1988, the New Jersey Department of Environmental Protection (New Jersey DEP) adopted permanent regulations that fill some of the regulatory gaps in CAFRA. For example, many developers avoided the CAFRA restrictions by keeping their projects below the 25-unit threshold. With its finer mesh net, however, the new rules scoop up many of the smaller projects that slipped through CAFRA.

In 1989, a bill that would establish a coastal commission was introduced in the state senate. The proposed coastal commission would develop a comprehensive plan to protect coastal resources, including wetlands, and manage development in the CAFRA zone. One of the goals of the commission would be to increase the total acreage of viable wetlands. The 17-member commission would be comprised of commissioners from several state agencies, such as DEP, and governor-appointed representatives from the building, fishing, and tourism industries and from environmental groups.

FWPA is designed to be tough enough to meet federal standards for delegating primary authority for wetlands regulation from the Corps to the state. The act requires the state to "take all appropriate action" to secure assumption of the permit jurisdiction exercised by the Corps.[10] For consistency, the act incorporates many of the same features and definitions of the federal 404 program; designation of freshwater wetlands, for instance, is based on hydrology, soils, and vegetation, just like the 404 program.

Both laws exempt certain agricultural activities. FWPA exempts many of the same activities as the federal program, such as normal farming, silviculture, and ranching activities, while the Wetlands Act exempts existing commercial agricultural activities, such as production of salt hay.

Permitting

New Jersey has established three classes of freshwater wetlands—exceptional, ordinary, and intermediate—with different levels of building restrictions for each category. The classification is based on "resource value" (that is, a wetland's value for such things as wildlife habitat as opposed to its economic value). The exceptional class includes wetlands that either discharge into waters where trout spawn or that provide habitat for endangered or threatened species. The requirements for development in this category are very strict, and, according to one New Jersey DEP official, "developers should not even bother trying to get a permit for exceptional wetlands." The ordinary class includes any isolated wetland that is smaller than 5,000 square feet and that is surrounded by development on at least 50 percent of its borders, as well as drainage ditches, swales, and stormwater detention facilities. The intermediate category includes every other type of wetland.

DEP also regulates certain activities (removal, excavation, disturbance, and so on) in a buffer or transition zone around exceptional and intermediate wetlands. The transition zone is 75 to 150 feet for exceptional wetlands and 25 to 50 feet for wetlands in the intermediate class. If, however, an applicant can prove that the proposed activity will have no substantial impact on the adjacent freshwater wetland, or that prohibition of the activity will cause undue hardship, then DEP may grant a waiver to reduce the size of the transition zone—but not below the minimum widths of 75 or 25 feet.

FWPA incorporates many of the provisions of EPA's Section 404(b)(1) guidelines, such as the water dependency and practicable alternative tests. For instance, permits shall not be issued for nonwater-dependent activities unless no practicable alternatives can be found, or for activities that will jeopardize endangered species, adversely affect ground or water quality, or are not in the public interest. New Jersey's definition of practicable alternatives is almost identical to EPA's.

FWPA established a set of rebuttable presumptions that a practicable alternative exists for any nonwater-dependent activity, and that such an alternative would have a less adverse impact on the aquatic environment. In order to rebut these presumptions, the applicant must demonstrate: one, that the proposed project cannot reasonably be accommodated on other sites in the general region; two, that the basic purpose of the project cannot be accomplished if the project is reduced in size, scope, configuration, or density; and three, that the applicant has made a reasonable attempt to overcome any constraints, such as zoning or infrastructure, that may have precluded the use of alternative sites. For development proposals in wetlands classified as exceptional, the applicant must also demonstrate that the proposed project will meet a compelling public need or that denial of a permit would impose an extraordinary hardship on the applicant. The rebuttable presumptions pose significant hurdles for developers to overcome. According to one New Jersey developer, in order to get a permit, you have to jump through several hoops that get progressively smaller. The stringent criteria essentially limit development in exceptional wetlands to public projects such as road construction.

New Jersey does not have a formal joint permit process with the Corps, but it does regularly review and comment on pending projects with the Corps and federal resource agencies.

The criteria for developing in coastal or tidal wetlands resemble those that apply to development in freshwater wetlands. The Wetlands Act of 1970 does, however, make no distinction among wetlands types (there is no classification scheme), nor does it require regulation of a buffer zone' although DEP, through CAFRA, does require buffers around coastal wetlands for projects needing a CAFRA permit. DEP prohibits development in coastal wetlands unless a proposed project will fulfill four criteria: it is water dependent; it has no prudent alternatives on a nonwetland site; it will result in a minimal alteration of natural tidal circulation; and it will result in minimal alteration of the natural contour or the natural vegetation of a wetland.

Mitigation

In general, DEP requires a 2:1 mitigation ratio for both coastal and freshwater wetlands, although it allows ratios as low as 1:1 under certain circumstances. The mitigation must be performed prior to, or concurrent with, activities that will permanently disturb wetlands, and immediately after activities that will temporarily disturb them. New Jersey distinguishes between four types of mitigation—restoration, creation, enhancement, and contribution—with a different ratio of required mitigation for each. For restoration, which includes "actions taken to recreate wetlands characteristics, functions, and habitat" that were affected by construction, and for creation, the ratio is 1:1—although, according to one DEP official, 2:1 is usually required. For enhancement, however, the ratio is 7:1. One reason why creation, which is generally perceived as the riskiest of all mitigation options, is only 2:1 whereas enhancement is 7:1 is that creation is much more expensive. In most states (though not in Florida) it is the other way around, with creation commanding the highest ratio. After all other mitigation options have been exhausted, applicants may donate land or money to a mitigation bank, which is not yet open for business.

DEP may require creation or restoration of a freshwater wetland of "equal ecological value" to the wetland that was lost. Equal ecological value is defined as "functional equivalency, including similar wildlife habitat, similar vegetative species coverage and density, [and] equivalent flood water storage capacity;"[11] however, no standards or guidelines have yet been established. DEP demands on-site mitigation where possible, but, after consultation with EPA, DEP may allow off-site mitigation if a developer either guarantees that no future development will occur on the mitigation site or makes a contribution to the wetlands mitigation bank. Contributions to the mitigation bank are based on the cost of purchasing and either creating or restoring a freshwater wetland of equal ecological value.

The mitigation bank is governed by a seven-member mitigation council composed of the governor and six members appointed by the governor with consent of the senate—two from the building and development industry, two from environmental groups, and two from institutions of higher learning. The council, which is responsible for using the funds of the mitigation bank to finance mitigation projects, is authorized to buy degraded wetlands that can be restored and to preserve critical wetlands and transition or buffer areas. In 1989, the council asked leading members of the development community if they would be interested in establishing a mitigation bank. Developers balked, however, at setting up a mitigation bank without some assurance that DEP would allow development in wetlands.

With their permit applications, developers must submit a mitigation proposal that includes:
- a monitoring and maintenance plan to ensure 85 percent survival of wetlands plants and 85 percent plant coverage for at least three years;
- a schedule of completion; and
- a description of the site's proposed hydrology.

Enforcement

For illegal activities in freshwater wetlands, DEP may assess civil penalties of up to $10,000 per day of violation, and up to $25,000 per day for those who willfully or negligently violate the act. In the first year of FWPA's operation, two administrative penalties were assessed, both for $250,000. In coastal wetlands, violators may be fined up to $1,000 per day and are liable for the cost of restoring the wetland to its prior condition. In 1988, 85 administrative penalties were assessed for a total of $57,650 for both coastal programs—CAFRA and the Wetlands Act of 1970.

Comments

New Jersey's wetlands programs stand out from those adopted by other states. In such a heavily industrialized, densely populated state, it is remarkable that so many wetlands still remain. New Jersey's classification of freshwater wetlands is unique. Like some SAMPs, New Jersey's freshwater wetlands program essentially classifies wetlands as suitable or unsuitable for development and establishes requirements for different types of mitigation. The classification puts developers on notice that exceptional wetlands are off-limits to development unless unusual circumstances exist. Developers have already felt the impact of the fledgling freshwater wetlands program. According to one New Jersey DEP official, most developers either steer clear of wetlands or keep their fill to an absolute minimum in order to qualify for a general permit.

Although it is not yet active, the mitigation bank also sets New Jersey apart from most other states. And if the proposed coastal commission is ever established, New Jersey will join the small club of states who have such commissions. The commission could streamline the permit process for the coastal area by providing "one-stop shopping;" alternatively, as its critics contend, the commission could just add another layer of bureaucracy to the development process.

California:
Two Programs for the Coast

California has always thought of itself as a trendsetter. Besides Oregon, it was the only state with a coastal zone management program before 1972, the year in which the federal Coastal Zone Management Act was enacted. With over 1,100 miles of varied coastline, it is not surprising that California took the lead in regulating coastal activities. Despite heavy losses of coastal wetlands, primarily due to agricultural conversion, California still possesses nearly 90,000 acres of salt marshes, more than four times the acreage of salt marshes in Oregon and Washington combined. Unlike several other coastal states, California has not adopted laws specifically to protect either its coastal wetlands or what remains of its inland wetlands. Instead, it strictly regulates activities within the coastal zone, including development in coastal wetlands, under the state's federally approved Coastal Zone Management Program, and leaves control of inland wetlands to the Corps.

California is unusual in that two separate agencies with different enabling legislation monitor distinct geographical regions of its coastal zone. The San Francisco Bay Conservation and Development Commission (BCDC) regulates activities along the coast of the San Francisco Bay while the California Coastal Commission (CCC) covers the rest of coastal California. BCDC was established as a temporary agency in 1965 and made permanent in 1969 following enactment of the McAteer-Petris Act. Likewise, CCC, which was modeled after BCDC, was established temporarily in 1972 and permanently in 1976 with enactment of the California Coastal Act. In 1977, both coastal programs were among the first state programs approved by the U.S. Department of Commerce under the federal Coastal Zone Management Program.[12]

BCDC was set up primarily to regulate uncontrolled filling of the Bay and to increase public access to the shoreline. It has permit authority over development within the Bay and, with few exceptions, issues permits for water-dependent uses only. BCDC's jurisdiction includes all of San Francisco Bay, certain creeks and sloughs that are part of the Bay estuary, salt ponds, and

Development abuts a salt marsh in the San Francisco Bay. Although regulated by BCDC and the Corps, these wetlands often receive polluted runoff from roads and parking lots.

some other areas that have been diked off from the Bay and are now used for duck clubs, game refuges, or agriculture. All other wetlands located behind existing dikes are excluded from its jurisdiction; and according to Robert Batha, environmental planner for BCDC, "People can do just about whatever they want behind the dikes as long as they provide public access to the beach."[13]

Unlike the limited jurisdiction of BCDC, CCC exercises sweeping authority over virtually all types of development along the California coast within 1,000 yards of the shore, and is the principal agency involved in regulating development in the coastal zone, including development in wetlands and associated habitat areas located in this zone. Guided by the 1976 California Coastal Act, which directs the commission to limit development in wetlands by encouraging development in more urbanized areas, CCC regulates a broad range of activities, from enlarging a house to siting industrial facilities and ports. "In many ways," notes one observer, "the CCC is not so much a coastal management agency as it is a state comprehensive land use planning and regulatory agency that has jurisdiction over some of this nation's most complex urban problems, fastest growing areas, and most expensive real estate that just happens to be along a coastline."[14]

Initially, CCC sought to create a comprehensive plan for the state's coastal zone and to regulate coastal development through a permit process. CCC retains authority over development in the coastal zone until local governments adopt coastal plans and ordinances that are consistent with the California Coastal Act. Once CCC certifies the plan, the local government can assume its own permitting responsibilities. By 1989, nearly all local plans were approved by the commission.

When it established CCC in 1976, the California legislature also created a nonregulatory agency, the Coastal Conservancy, to resolve conflicts over use of the state's coastal resources. A project-oriented agency with no permitting authority, the conservancy is authorized to buy land, shift development rights, improve beach access, and restore coastal resources such as sand dunes and wetlands. Typical projects include revitalizing aging waterfronts, restoring diked, tidal wetlands, purchasing shorefront property for parks and open space, and building public access ramps to the beach.

A History of Wetlands Losses

California has lost a large proportion of its wetlands. In the central valley alone, some experts estimate that over 90 percent of the freshwater wetlands were destroyed when most of the valley was converted to farmland. Coastal wetlands fared little better. San Francisco Bay alone has lost over 80 percent of the wetlands that it contained 100 years ago, leaving little more than 100 square miles of tidal marsh undiked.[15] Outside San Francisco Bay, California once contained over 300,000 acres of coastal wetlands; today, only about 80,000 acres remain.[16] In addition to those wetlands lost, many were seriously altered through sedimentation and/or filling. Although it is not their primary purpose, both BCDC and CCC have helped to stem the tide of wetlands losses within the coastal zone by controlling development in wetlands and requiring creation of new wetlands or restoration of old ones when filling occurs.

Permitting

The California Environmental Quality Act (CEQA) of 1971 plays an important role in virtually every development in the state and particularly in its coastal zone. Under the act, a local government agency reviews a development proposal to determine if a proposed project is subject to CEQA. Certain activities, such as repair and maintenance of existing facilities, are categorically exempt. If a project is not exempt, then the local or "lead" agency will conduct an "Initial Study" to determine whether or not the project will have a significant impact on the environment. If it will not, then the developer can breathe a sigh of relief. If it will, then the agency must prepare an environmental impact report (EIR), which is not unlike an EIS under the National Environmental Policy Act. The EIR evaluates all the possible environmental impacts of a proposed development, determines if feasible, less environmentally damaging alternatives exist, and develops possible measures that would mitigate adverse impacts. Under CEQA, a public agency cannot approve a project if development on an alternative site could substantially lessen the environmental impacts or without incorporating mitigation measures that would have the same effect. Typically, the coastal agencies begin their review of a project after the EIR is complete—a policy that irks most developers who feel that they have already conducted sufficient environmental reviews and who have often made substantial changes to their original plans in order to satisfy the regulatory agencies.

With few exceptions, both agencies will allow wetlands filling only for water-dependent projects, such as ports, water-related industries, bridges, and water-oriented recreation, and only if no practical alternatives exist or if a project will increase public access to the shore. BCDC will also approve projects that improve shoreline appearance. Generally, both agencies conduct a public interest review to determine whether or not the public benefits of a proposed project clearly outweigh the public detriments. Any project that does not show a net benefit to the public will be rejected. In addition, CCC will not approve projects that jeopardize the "functional capacity" of a wetland, which is defined as the ability of a wetland or estuary to be self-sustaining and to maintain a diversity of species. In order to protect environmentally sensitive areas from off-site development impacts, CCC further requires that a project include buffers to separate such areas from development. The commission's goal is to create a "no-man's-land" that shields sensitive environments from urban development. Development within buffer areas is limited to access paths, fences, and similar uses that have either beneficial effects or at least no significant adverse impacts. Buffer width varies with each project, but is usually at least 100 feet.

BCDC's permits fall into two categories: minor projects, such as routine maintenance dredging, installation or repair of stormwater or wastewater outfall pipes, or levee stabilization work; and major projects, like expanding a boat-storage facility, constructing a hotel at the site of an abandoned warehouse, constructing a marina, or installing riprap and bulkheads for shoreline protection. Unless problems are encountered, the time taken to obtain a permit for a minor project varies from one to two months, and for a major project is about three months, although it sometimes takes longer. BCDC boasts of administering one of the speediest regulatory programs in the state. Under the provisions of the McAteer-Petris Act, if BCDC fails to act within 90 days of the filing of a permit application, the permit will be automatically granted. Although the commission occasionally fails to meet the deadline, no permits have been automatically issued. According to BCDC, a delay in processing an application usually reflects the commission's objection to the proposed project, and an extension is offered to grant the applicant additional time to modify the proposal so that it may ultimately be acceptable to the commission.

BCDC receives roughly 20 to 30 applications each year for major projects and about 100 to 140 for minor projects. It approves the great majority of projects, although all approvals include conditions to ensure that any adverse impacts on the Bay are minimized. (Likewise, the San Francisco District Office of the Corps issues permits for over 98 percent of the applications it receives, although two-thirds of the applications are modified before being approved.) Between 1970 and 1988, the commission granted 362 major and 1,938 minor permits and denied only 18 and 10 respectively (see Figure 3.2). BCDC has been in existence for over 20 years, and by now most people know not to bother applying for a permit to fill Bay wetlands for nonwater-dependent projects.

Scarcity of Mitigation Sites

In 1973, BCDC first required mitigation for Bay fill. Since then, it has substantially improved public access and significantly reduced wetlands filling. From 1974 to 1988, the commission allowed about 265 acres of wetlands to be filled and, in return, required creation or restoration of 1,134 acres, for an overall mitigation ratio in excess of 4:1.

Both BCDC and CCC rely on mitigation to offset adverse impacts of development. Mitigation cannot, however, make a project acceptable if it does not meet all the requirements of the McAteer-Petris Act or the California Coastal Act. Mitigation typically involves either restoring diked wetlands by breaching the dikes to reopen the area to tidal action or enhancing existing tidal marshes by removing debris and constructing channels to improve water circulation and drainage.

Although CCC typically requires mitigation at a 1:1 ratio, it can demand up to 4:1, and some developers do more. One commission official noted that he had not seen anything less than 1:1, and, in an extreme case, one public project's mitigation-to-fill ratio was 400:1.

FIGURE 3.2
19-YEAR SUMMARY OF PERMITS, FILL, AND MITIGATION IN SAN FRANCISCO BAY

Year	Major Permits Granted—Denied		Minor Permits Granted—Denied		Fill Authorized (Acres)	Mitigation Required (Acres)	Net Change (Acres)
1970	12	1	66	0	72.0	0.0	−72.0
1971	26	4	61	0	25.1	0.0	−25.1
1972	12	3	80	0	7.0	0.0	−7.0
1973	17	1	71	0	4.4	0.0	−4.4
1974	20	0	107	1	83.0	357.0	+274.0
1975	10	0	87	0	0.0	5.0	+5.0
1976	14	0	110	0	2.2	0.0	−2.2
1977	20	0	116	0	27.7	44.5	+16.8
1978	23	1	104	4	7.8	5.9	−1.9
1979	34	0	120	2	17.8	21.2	+3.4
1980	19	1	105	1	25.4	55.4	+30.0
1981	23	0	134	0	5.1	49.6	+44.5
1982	26	0	104	0	24.0	286.0	+262.0
1983	23	0	105	0	4.0	9.0	+5.0
1984	15	3	135	0	17.0	29.0	+12.0
1985	15	1	98	0	30.0	90.0	+60.0
1986	20	0	108	0	11.4	22.4	+11.0
1987	16	2	108	0	5.4	3.4	−2.0
1988	17	1	119	2	4.7	156.9	+151.3
Totals	362	18	1,938	10	374.0	1,134.4	+760.4

Note: Some authorized projects have not been built, and some projects may have been changed pursuant to amendments to permits.
Source: San Francisco Bay Conservation and Development Commission, "1988 Annual Report," issued January 1, 1989.

Although BCDC's policy is that mitigation should be at least 1:1, it has not established a specific mitigation ratio or formula. Instead, it prefers to fix mitigation requirements on a case-by-case basis, assembling an individual mitigation package for each project. One consultant in San Francisco stated that BCDC usually requires 2:1 mitigation and occasionally 4:1, but can require as much as 10:1. BCDC prefers in-kind, on-site mitigation but has approved projects involving off-site or out-of-kind mitigation. Both agencies generally stipulate that mitigation must occur prior to or concurrently with development, and the agencies suggest that developers submit restoration and management plans with their permit applications. The CCC interpretive guidelines state that, "at a minimum, the permit will be conditioned to assure that restoration will occur simultaneously with project construction."[17] Moreover, restoration must occur within the same general region (within the same stream, lake, or estuary) where the fill occurred.

Until recently, CCC allowed previously diked wetlands to be converted back to a tidal marsh in order to meet its requirements for mitigation. Now, though, CCC recognizes that these wetlands provide valuable wildlife habitat, albeit often only seasonally, and developers must create wetlands out of upland to compensate for altering coastal wetlands. With fewer sites to choose from, finding suitable mitigation sites has become a never-ending quest for some developers. Most areas around the Bay are either protected, already developed, or very expensive. South San Francisco Bay is currently experiencing extreme development pressure, and an acre of diked wetlands can cost $300,000. A study commissioned by BCDC found that five of 14 projects reviewed were delayed because the permittee was unable either to find or to acquire a suitable restoration site.[18] As a result of this dearth, developers have looked towards less expensive upland sites in the North Bay. Many of these relatively cheap sites are former wetlands that were diked years ago and converted to agriculture; changing them back to wetlands is often as easy as breaching a dike to return tidal flow. A few developers have purchased or taken options to buy these diked wetlands as possible mitigation sites for future projects.

According to an attorney in San Francisco, since 1984, BCDC has operated a de facto no-net-loss policy that effectively rules out wetlands enhancement. Given the scarcity of suitable mitigation sites around the Bay and BCDC's antipathy to enhancement, it is virtually impossible to obtain approval to fill wetlands in the Bay. BCDC contends that it has no policy forbidding wetlands enhancement, but it simply recognizes that other agencies will likely disapprove any development involving wetlands enhancement as mitigation for fill. With the exception of the Oakland Airport expansion, which is a public project, no sizable wetlands development has been approved in Bay wetlands since 1984.

Monitoring and Enforcement

Both BCDC and CCC lack the resources to conduct any significant degree of on-site monitoring. (Perhaps this is one reason why, according to an official with CCC, mitigation has not been very successful in California.) Instead, their staffs focus on investigating and prosecuting unauthorized fills. CCC did set up a system for "self-monitoring" by professionals, and the commission requires annual reports, prepared and submitted by independent consultants, on the success or status of each mitigation project. However, a review of 14 BCDC-permitted projects revealed that in only three cases was the permittee required to monitor the restoration (only one had actually done any monitoring) and that in only two cases was the permittee required to maintain the stipulated improvements to ensure that the restoration project functioned as designed.

Like monitoring, enforcement receives short shrift—chiefly because of the limited resources of both BCDC and CCC. Both commissions can in theory issue cease and desist orders, require corrective measures, and impose civil penalties. BCDC may issue penalties of between $10 and $1,000 per day for a violation of either the McAteer-Petris Act or a permit condition, with a total maximum penalty of $20,000. In 1988, four penalties totalling $8,700 were imposed. According to BCDC, in practice, enforcement cases are resolved satisfactorily through voluntary compliance or by issuing or amending a commission permit.[19]

Different Roles, Different Reputations

By most accounts, BCDC has been very successful in its limited mission and jurisdiction. By requiring developers to reopen diked areas to tidal action as a condition for approving new Bay fill, the commission has enlarged the open waters of San Francisco Bay and increased the amount of Bay shoreline open to public access from only four miles to over 100 miles.[20] BCDC is well respected by both developers and conservationists as an even-handed and efficient keeper of the Bay's resources.

By contrast, CCC has become a controversial and often unpopular agency. Bills are frequently brought before the state legislature to abolish, or at to least to cripple, the agency. Environmentalists accuse CCC of bowing too often to pressure from developers; local governments resent a state commission meddling in local affairs; and developers generally find the commission's requirements both unpredictable and excessive. Given the commission's Herculean task of regulating development along the entire California coast (save for San Francisco Bay), such unpopularity is hardly surprising. In its attempt to balance various competing interests, CCC has upset almost everybody at one time or another, which may be one indication that the agency is doing its job well.

Besides BCDC and CCC, two other state agencies play important roles in regulating development in wetlands: the State Department of Fish and Game, and the State Water Resources Control Board. The Department of Fish and Game comments on applications for both state and federal permits, and the water resources board, which is responsible for protecting the quality of certain waters in the Bay area, reviews and certifies Corps permits for compliance with state water-quality standards. Both state agencies have adopted a no-net-loss policy for wetlands.

Oregon:
Emphasis on the Estuaries

The first state to adopt a state planning law and one of the first to protect its coastal wetlands, Oregon's wetlands program is very similar to the federal program. Indeed, Oregon initially seriously considered assuming the federal Section 404 program, but decided that it was not worth the effort, even after EPA changed its rules to make it easier for states to do so. According to a state official, "EPA's rule change did not change anything"—the Corps would still have retained authority over navigable waterways and adjacent wetlands, which is where most permit problems in Oregon occur. For example, in the first quarter of fiscal year 1988, nearly 90 percent of the permit actions delayed more than 60 days were in those areas where the Corps would have retained its authority.[21]

Under Oregon's Removal-Fill Law (O.R.S. 541.605–541.695 and 541.990, September 1985), a permit is required from the Division of State Lands (DSL) to fill or remove material from "waters of the state," which include all tidal and nontidal bays, streams, lakes, and "other bodies of water in this state, navigable and nonnavigable." The law applies to removals or fills of 50 cubic yards or greater and covers activities in both coastal and freshwater wetlands, including isolated wetlands.

Permitting

Like the federal program, Oregon's program prohibits the issuance of a permit unless: one, no "practicable alternative" exists to the proposed fill that would have less adverse impact; two, the project will not cause "significant degradation of municipal water supplies, aquatic life and habitats;" and, three, "appropriate and practicable" steps have been taken to minimize adverse impacts. In addition, DSL is required to issue a permit to remove material from a wetland if the removal will not be "inconsistent with the protection, conservation and best use of the water resources of this state. . . ."[22]

Once DSL receives an application, it issues a public notice, reviews any comments it receives, and decides whether or not to issue a permit according to such factors as the environmental and economic consequences of the proposed removal or fill, its effect on health, safety, and welfare, and on water quality and quantity. Before a permit is issued, DSL must determine that the proposed fill will be consistent with state-acknowledged local comprehensive plans.

As in the federal program, the state can also issue after-the-fact permits (although, according to one DSL official, such permits are uncommon). After-the-fact permits are only considered if: the project has caused no harm to the state's water resources; the project has been completed in its entirety; the responsible person had no knowledge of the permit requirements; and the project meets other permit requirements. DSL strongly discourages nonwater-dependent activities and permits them only if the fill (the rules do not mention nonwater-dependent removals) is for a public use and if the public need outweighs the harm to navigation, recreation, and fisheries. The state can also close off certain wetlands areas to development in advance of a permit request. But it cannot classify or designate areas as suitable for development, as EPA can under the Clean Water Act.

The state receives roughly 1,000 applications per year and approves about 98 percent of them. Most applications are for small fills, such as routine pipeline crossings and placement of riprap. All applications have conditions attached to them by the state to minimize adverse impacts. A permit application normally takes between 45 and 60 days to process, although permits for problem sites can take 90 days or more. DSL and the Corps process permits jointly, which helps avoid duplication. A local consultant commented that whereas the state processes permits fairly quickly, usually within 60 to 90 days, the Corps is notoriously slow and often delays a project for several months. DSL does not recognize the Corps' nationwide permits and still requires an applicant to obtain a state permit for areas that are covered by a Corps nationwide permit.

Mitigation

Although the permit program applies to both tidal and nontidal wetlands, the mitigation requirements for tidal fills are more stringent, and the standards are more explicit. Oregon, like other coastal states, guards its tidal wetlands more carefully than nontidal wetlands. The state usually only requires mitigation for "significant fills" in nontidal wetlands, but it has not established clear standards or guidelines for the amount of mitigation required. According to one consultant, mitigation for development in nontidal wetlands is generally determined on a 1:1 basis, although the trend is towards requiring more than 1:1. DSL enjoys broad discretion in the issuance of permits for development in nontidal wetlands and in the requirement of mitigation.

For projects in tidal or intertidal wetlands, each application for fill or removal must be accompanied by a mitigation plan. Under certain circumstances, DSL can waive the mitigation requirements—for example, if the mitigation is unfeasible or if the economic and public benefits outweigh the potential degradation to the estuary. Mitigation must occur in the same estuary as the removal or fill.

DSL encourages, but does not require, in-kind mitigation. Its rules for estuarine mitigation state that mitigation shall maintain diversity of habitat and species; however, the law does not mandate that every habitat and species be replicated by the proposed mitigation. In fact, DSL often requires a different type of vegetation in mitigation projects than existed in the wetland that was filled. For example, although the state regulates development in reed canary grass mashes and requires mitigation when such mashes are filled, it generally demands that other types of wetlands be created in their place.

In estuarine areas, the amount of mitigation required will be proportionate to the relative value of the habitats involved. The state evaluated the various habitat types found in its estuaries in terms of biological productivity and species diversity and devised a set of relative values from a scale of one to six for each type; for instance, a marine or brackish seagrass marsh is assigned a value of six, but an unvegetated wetland receives only a two or three. Oregon has adopted a no-net-loss estuarine mitigation policy: the mitigation ratio will be at least 1:1 and in some cases as high as 6:1. The values, shown in Figure 3.3, are fed into an equation to determine how much area of one habitat is needed to mitigate each acre of another habitat lost to removal or fill. Under certain circumstances, the director of DSL may adjust the relative values, up or down, by up to 25 percent. The equation is as follows:

$$AM = (RVd/RVm)(AD),$$

where AM = area of mitigation site;
RVd = adjusted relative value of the development site;
RVm = adjusted relative value of the mitigation site;
and AD = area of development site.

The advantage of this system is that developers can compute, in advance, the amount of mitigation that will be required for a given fill or removal in a given area. It gives developers the predictability that they complain most wetlands programs lack. In addition, the system also provides economic disincentives to removals or fills in high-value estuarine areas.

Enforcement

Only seven DSL staff members monitor permitted projects for compliance. According to one DSL official, however, most projects are checked for compliance. An applicant must notify the state when the mitigation begins and must also submit an annual monitoring report for the first three to five years. One DSL official estimated that about 100 unpermitted fills or removals take place each year, but, because of the state's monitoring program, far fewer permit violations occur.

DSL may take a number of actions against those who violate the state removal-fill law. It can assess civil penalties and/or force the violator to remove the fill and restore the site. But, according to one state official, "about 80 percent of violations are small removals or fills from people who unwittingly violate the rules," and the state usually treats such people leniently and prefers to negotiate a voluntary settlement where the violator will rectify the damage done.

DSL, which can impose civil penalties of up to $10,000 per day of violation, uses the following formula to compute the amount of the civil penalty:

$$\text{Penalty} = \$B(V \times P \times C \times I).$$

B is a base fine, equal to the permit application fee. V represents the wetland's recreational, biological, navigational, or fisheries value. Wetlands with a minor resource value, including wetlands with a history of dis-

FIGURE 3.3
RELATIVE VALUES[1] OF SELECTED ESTUARINE HABITAT TYPES IN OREGON

OREGON ESTUARIES (EXCEPT THE COLUMBIA RIVER)

GENERALIZED SUBSTRATE CHARACTERISTICS	SUBTIDAL HABITATS			INTERTIDAL HABITATS						SALINITY REGIME
	UNVEGETATED	ALGAE	SEAGRASSES	UNVEGETATED	ALGAE	SEAGRASSES	LOW MARSH	HIGH MARSH	FORESTED WETLAND	
ROCKY–BEDROCK (Max. Grain Size >256 mm)	1.0	2.0	–	1.0	2.0	–	–	–	–	FRESH
	2.0	3.0	–	2.0	3.0	–	–	–	–	BRACKISH
	2.0	3.0	–	2.0	3.0	–	–	–	–	MARINE
COBBLE–GRAVEL (Grain Sizes from 1.0 mm to 256 mm)	1.0	2.0	4.0	1.0	2.0	4.0	4.0	3.0	3.0	FRESH
	2.0	3.0	6.0	2.0	3.0	6.0	5.0	4.0	–	BRACKISH
	2.0	3.0	6.0	2.0	3.0	6.0	5.0	4.0	–	MARINE
SAND (75% Grain Sizes from 0.0625 mm to 1.0 mm)	2.0	3.0	4.0	2.0	3.0	4.0	4.0	3.0	3.0	FRESH
	3.0	4.0	6.0	3.0	4.0	6.0	5.0	4.0	–	BRACKISH
	3.0	4.0	6.0	3.0	4.0	6.0	5.0	4.0	–	MARINE
SANDY–MUD	2.0	3.0	4.0	2.0	3.0	4.0	4.0	3.0	2.0	FRESH
	3.0	4.0	6.0	3.0	4.0	6.0	5.0	4.0	–	BRACKISH
	3.0	4.0	6.0	3.0	4.0	6.0	5.0	4.0	–	MARINE
MUD (75% Grain Sizes < 0.0625 mm)	2.0	3.0	4.0	2.0	3.0	4.0	4.0	3.0	3.0	FRESH
	3.0	4.0	6.0	3.0	4.0	6.0	5.0	4.0	–	BRACKISH
	3.0	4.0	6.0	3.0	4.0	6.0	5.0	4.0	–	MARINE

1. Relative values are based on natural biological productivity and species diversity of specific habit types. A "–" means the habitat type probably does not exist.

Source: Oregon Division of State Lands, *Estuarine Mitigation: The Oregon Process* (Salem, Oregon: author, April 1984), p. 16.

turbances or physical alterations, are assigned a value of one. Those with an average resource value are given a three, while wetlands with major resource value, for example, those that contain salmon or trout spawning areas or important waterfowl habitat, are given a four. P represents a violator's prior knowledge of the removal and fill law. For those with no prior knowledge, the figure is one; for those with prior knowledge, three; and for those who previously violated the removal and fill law, five. C, the cooperation value, represents a violator's efforts to correct the violation. It ranges from 0.5 for those who are cooperative to 2.0 for the uncooperative. I, the impact value, is a measure of the impact that the removal or fill will have on water resources. For minor impacts, the value is two; for average impacts, three; and for major impacts, four.

Miitigation Banking

In 1987, the state enacted a Wetlands Mitigation Banking Act, which authorized DSL to establish four pilot mitigation banks by July 1991. The banks will be used only for relatively small removals or fills—five acres or less—and only after all practical on-site mitigation methods have been exhausted. DSL is developing rules and criteria for selecting sites and for managing and evaluating the banks. As of December 1989, no pilot mitigation banks had been established, although a mitigation bank was previously established in Astoria, on the northern coast of Oregon.

Galien River as it winds through a 400-acre wetland before entering Lake Michigan at New Buffalo, Michigan. In presettlement times, this wetland was a 90-foot-deep lake called Lake Potawatomia, which sediments from upstream land uses filled completely.

Michigan:
Running Its Own 404 Program

Swathed by the Great Lakes and their glaciated wetlands, Michigan contains over 2 million acres of wetlands. Most are protected by Michigan's wetlands program, which combines comprehensive wetlands protection with stiff penalties for violations. Michigan's wetlands fall under the jurisdiction of a variety of state laws such as the Inland Lakes and Streams Act, the Shorelands Protection and Management Act, and the Great Lakes Submerged Lands Act. But the centerpiece of Michigan's wetlands management program is the Goemaere-Anderson Wetland Protection Act of 1979[23] (referred to hereafter as the Wetlands Act). The Wetlands Act fills in the gaps left by the other programs. For example, the Inland Lakes and Streams Act regulates certain activities in inland lakes and streams and their associated wetlands *below* the ordinary high-water mark, and the Wetlands Act covers activities *above* the mark. Michigan is especially strict about development in its wetlands along the Great Lakes and Lake St. Clair. One state official remarked that developers should not even think about developing in those wetlands. Thus far, Michigan is the only state with both a sufficiently strict program and the political inclination to assume administration of the federal Section 404 permit program within its borders (which it did in 1984).

Although Michigan's wetlands program is broader than the 404 program, the two have much in common. For example, both contain similar definitions of wetlands and permitting criteria, and Michigan also issues five-year general permits on a state or county basis for activities that are similar in nature and will have only minimal individual and cumulative adverse environmental impact. Michigan, like several other states, also has an environmental protection act that requires all individuals and organizations, public or private, to prevent or to minimize adverse environmental impacts of their activities on the state's natural resources, including wetlands.[24]

Activities Regulated

The Wetlands Act splits wetlands into two broad categories—contiguous and noncontiguous—with different regulations for each group. Contiguous wetlands lie along lakes, streams, or ponds. Michigan regulates all of these wetlands, regardless of size. Noncontiguous wetlands, isolated either hydrologically or geographically from surface waters, are defined as wetlands located more than 1,000 feet from the ordinary high-water mark of one of the Great Lakes or Lake St. Clair, or more than 500 feet from the ordinary high-water mark of an inland lake, pond, river, or stream. Noncontiguous or isolated wetlands smaller than five acres are exempt from regulation, unless the Department of Natural Resources (DNR) determines that such a wetland is essential to protect natural resources from pollution, impairment, or destruction. Moreover, DNR may regulate wetlands that are otherwise exempt if the wetland supports federal endangered species or plants or animals of an "identified regional importance" or provides groundwater recharge. In counties with fewer than 100,000 people, noncontiguous wetlands are not regulated until a wetlands inventory is completed.

The Wetlands Act prohibits the following activities in wetlands without a permit from DNR: dredging, filling, or removing soil or minerals; draining surface water; and constructing, operating, or maintaining any use or development. Exempted activities include farming, ranching, logging, fishing, and trapping.

Michigan relies primarily on vegetation and aquatic life to delineate wetlands from uplands. However, in the absence of visible evidence of water, DNR will rely on soil types for wetlands identification. Michigan, like several other states, is in the process of mapping all of its wetlands.

Permitting

Since Michigan is essentially running its own 404 program, permit applications must go through a type of public interest review similar to that which the Corps performs and must meet similar tests to those which EPA has established, such as the water dependency and practicable alternatives test. In other words, in Michigan, a permit to develop in wetlands will be issued only if the activity is in the public interest, is "primarily dependent on being located in the wetland," or if no "feasible and prudent" alternative exists. Michigan goes a step further, however, and also considers the amount of wetlands remaining in an area and the cumulative impacts of the proposed project on wetlands in a particular watershed before it will issue a permit.

Once DNR receives a completed application, it has only 90 days to reach a decision; otherwise, by law, the project will be automatically approved.[25] DNR fails to meet the 90-day limit only occasionally. The statutes state that DNR may hold a public hearing within 60 days of receiving a completed application in order to allow the public to respond to the proposal. Usually, DNR holds hearings only for large projects or for small, controversial projects. Minor activities that qualify for a general permit normally do not merit public hearings unless DNR receives a request for a hearing within a 20-day notice period.

DNR formerly encouraged preapplication consultations at which developers and agency officials would meet to discuss a proposed project. DNR, however, discovered that most developers ignored its recommendations, and so the consultations were abandoned.

In general, if an applicant makes a reasonable attempt to protect wetlands, DNR will issue a permit. One DNR official estimated that in the southeastern Michigan district roughly 20 to 25 percent of applications are granted as applied for, about 25 percent are granted after the project design has been modified to reduce wetlands impacts, and about 50 percent are rejected. A large share of the rejected applications comes from what DNR calls "first-time developers"—that is, developers who are unfamiliar with the state's wetlands laws and who have not dealt with DNR before. "A lot of these developers come in with the attitude that 'this is my land and I am going to fill every square inch of it'," commented a water-quality specialist with DNR. "These projects are denied outright." If denied a permit, developers have essentially two choices: either they can appeal the case to an administrative law judge in a contested case hearing, or they can redesign the project to reduce its adverse impacts and resubmit their application. Most developers cannot afford to have their projects held up in hearings, which can take a year, and will instead try to negotiate with DNR.

DNR is not the only obstacle to wetlands development in Michigan. A growing number of local governments have enacted wetlands ordinances that go farther than the state regulations and cover development not only in the wetland itself, but in a specified buffer zone around the wetland as well.

Mitigation

Michigan's wetlands mitigation policy is ambitious: the state will tolerate no net loss of wetlands; mitigation will be on site and should replace lost functional values; and mitigation must be done before or concurrently with development. But the policy is also flexible: if on-site mitigation is not practical, off-site mitigation is permitted as long as it is within the same watershed and municipality. In cases where it is "inappropriate and impractical" to mitigate wetlands impacts either on site or off site within the immediate vicinity, DNR will consider mitigation elsewhere. When applying for a permit from DNR, applicants must show that they have done all they can to avoid any wetlands losses and that it is practicable to replace lost wetlands values.

Developers in Michigan, especially those who have been stung by the wetlands regulations, are taking additional precautions before purchasing or developing a site. Often, a developer will hire a consultant to determine if a site contains wetlands; and if it does, the developer will either lower its purchase offer or search for another site.

Enforcement

Michigan has the authority to impose some of the stiffest penalties in the United States on those who violate its wetlands laws. The state may slap violators with a civil fine of up to $10,000 per day of violation. Those who willfully or recklessly violate the law may face a fine of up to $25,000 per day and one year in prison. A second willful violation is a felony, punishable by fines up to $50,000 per day or by two years in prison, or by both. In addition, a standard condition on permits makes the applicant responsible for the success of the mitigation for five years. If the wetland does not perform as planned, DNR can make the developer fix it.

Yet, though the penalties are strong, enforcement is weak. Like most states, Michigan simply does not have the resources to monitor every site, so it follows up on controversial projects only. For most projects, once a permit is granted, DNR never sees the project again. And, according to a DNR official, when DNR does catch someone, the courts do not take the wetlands laws seriously and allow most violators to escape with just a slap on the wrist.

Comments

Although EPA retains the right to review wetlands fill permits, it seldom looks over Michigan's shoulder. A memorandum of agreement between Michigan and EPA establishes criteria for when EPA will involve itself in state wetlands permitting. EPA may, for example, review a permit if an applicant proposes: either to relocate 500 feet or more or to enclose (in a pipe) over 100 feet of a stream; to disturb more than one acre of wetland; or to place 10,000 cubic yards or more of fill in a wetland. According to DNR, EPA usually concerns itself with "major" projects only, such as those projects involving 10,000 cubic yards or more of fill. Of course, you can fill a lot of wetlands with less than 10,000 cubic yards.

Unlike Oregon and New Jersey, Michigan has not authorized mitigation banking, which it views it as a risky, experimental technique. According to one DNR official, "We are not confident about mitigation banking; the jury is still out."

Massachusetts:
Regulating Wetlands under Home Rule

Massachusetts was the first state to adopt a wetlands protection law and also one of the first states with a federally approved coastal zone management program (1978). The Jones Act (1963) and the Hatch Act (1966) were enacted to regulate activities in coastal wetlands and freshwater wetlands respectively. In 1972, the two acts were consolidated in the Wetlands Protection Act. Massachusetts now has one of the strictest wetlands programs in the nation.

The state's program is unusual in that it establishes general performance standards for different types of resource areas. The act presumes that wetlands provide at least one of the following seven values: protection of 1) groundwater, 2) water supplies, 3) fisheries, and 4) land containing shellfish; and protection from 5) storms, 6) floods, and 7) pollution—and that these values are in the public interest.

The regulations identify four inland and 11 coastal resource areas that are subject to protection. The inland areas are: bordering vegetated wetlands (that is, wetlands that border creeks, streams, rivers, ponds, and lakes); land under waterways or water bodies; banks; and floodplains. Bordering vegetated wetlands are the most highly protected resource area. The coastal areas are: coastal beaches; barrier beaches; dunes; banks; salt marshes; land under salt ponds; rocky intertidal shores; land under the ocean; designated port areas; land containing shellfish; and banks of, or land under, the ocean, ponds, streams, rivers, lakes, or creeks that underlie anadromous or catadromous fish runs. The resource areas are separately and individually regulated. Each contains a specific statutory wetlands value or interest and a corresponding set of performance standards. The regulations define each resource area and state which of the seven state interests are significant in that area.

A permit is required for any activity that will remove, fill, dredge, or alter specified wetlands areas. Certain exemptions apply, such as mosquito control, maintenance of utility lines, and agricultural activities on land that is currently being farmed. The act also identifies "limited projects," which, while not exempt from the permit requirements, are not subject to the same performance standards as, say, dredging or filling. Such projects include farming, forestry, utility operations, road construction, and maintenance of boat launching ramps, docks, piers, and duck blinds. As in New Jersey, in Massachusetts the regulations also apply to activities within a buffer zone that extends 100 feet from every resource area except floodplains. No performance standards apply to the buffer zone, however.

Unlike other states that rely on both vegetation and soils to identify wetlands, vegetation is the sole criterion in Massachusetts. For example, a bordering vegetated wetland is an area where at least 50 percent of the plant community is wetlands plants.

Administration

Home rule is strong in Massachusetts, as it is throughout the Northeast, and 347 town conservation commissions implement the state's wetlands program. Local programs, however, must be approved by the state and must adopt state-developed regulations. Unlike most states outside the Northeast, Massachusetts does not directly involve its state government in the day-to-day permit reviews. The commissions issue permits, called orders of conditions, with appeals handled by the state's Department of Environmental Protection (DEP). The commissions handle most applications, called notices of intent; only about 10 percent of the applications are appealed to DEP. No two commissions are alike, with each reflecting its town's attitude toward development and wetlands protection. For example, in Cape Cod, where the local community strives to slow the pace of development, applicants encounter much more rigorous permit reviews than in towns that endeavor to attract development. In addition, cities and towns may enact their own wetlands ordinances, which are usually stricter than state law. About 90 towns have such laws, and developers can be denied a permit under a local ordinance even if a proposal meets the state's requirements.

Permitting

On average, the state receives about 6,000 permit applications per year and denies very few of them. In 1987, however, a boom year for development in Massachusetts, the state received about 10,000 applications. It usually takes about six weeks to process a permit, but for large projects with considerable opposition, the process can take much longer. Applicants must identify all resource areas within a proposed development site and either show how the proposed activity will meet the performance standards or how the presumptions of significance will be overcome. Proposals to alter bordering vegetated wetlands face the toughest regulations. The performance standards allow work to be undertaken in such wetlands only under very narrowly defined circumstances.

The application process differs slightly depending on the size of the proposed alteration. For alterations under 5,000 square feet, the permit process works as follows. An applicant files a notice of intent with the local conservation commission and DEP. Following receipt of a completed application, the commission schedules a public hearing within 21 days. Once the hearing is closed, the commission is supposed to reach a decision in 21 days, but it may take longer, and an applicant's only recourse against a sluggish commission is an appeal to DEP, which could make matters worse. Unlike Michigan, where a permit is automatically issued if the regulatory agency does not act within the statutory time limit, Massachusetts has no such provision and the process can drag on for years. Eventually, the commission or DEP will issue an order of conditions, which either grants or denies a permit.

Permits usually contain 12 standard conditions that specify, for example, that the work shall conform to the approved plans and that all fill must be clean. Some commissions tack on additional conditions. For example, the Woburn Conservation Commission usually attaches up to 20 additional conditions that range from stipulating riprap aprons on all culvert outlets to requiring developers to have an environmental scientist on site during the mitigation. If an applicant receives a permit, he or she must wait 10 working days before beginning work to allow time for objections to be voiced to the terms of the order. Anyone, including DEP, can appeal the conditions. If DEP appeals, it will eventually issue what is called a superseding order of conditions. Recognizing that time is money to developers, local commissions often use the threat of delay to squeeze concessions out of developers. If an applicant is denied a permit or objects to the permit conditions, he or she can appeal to DEP. But the appeal process, which is supposed to take only 70 days, can take considerably longer. In 1988, due to DEP's huge backlog of cases, applicants could expect appeals to take two to three years to process. According to one Massachusetts developer, the main problem with that state's wetlands program is that it is woefully understaffed, causing intolerable delays.

For applications to alter over 5,000 square feet, an applicant must first go through the motions of applying to the commission for a permit, which will be denied, and then appeal to DEP. An applicant may request an adjuratory hearing before DEP, and if still unsuccessful, an applicant's next recourse is to apply for a variance. The commissioner of DEP will issue a variance only if: an overriding public need justifies the project; no alternative sites exists; the proposed mitigation will adequately compensate for the alteration; or a variance is necessary to avoid a taking. The first criterion automatically excludes most private projects, except perhaps low-income or elderly housing, and generally only public projects, such as sewage-treatment plants or roads, qualify for a variance. In 1988, out of about 11,000 permit applications, only three variances were granted. "You'll have grandchildren before you get a variance," said one DEP official, "it takes forever."

The state's rationale for what amounts to a ban on development involving alteration of over 5,000 square feet is that it defined bordering vegetated wetlands conservatively—that is, as areas with over 50 percent of wetlands vegetation—whereas other states define wetlands more broadly as areas that support a "prevalence" of wetlands plants. This argument has not appeased builders and developers, however, who argue that the 5,000-foot threshold provides the "least bit of weasel room to allow a developer to use his site economically."

FIGURE 3.4
SUMMARY OF SIX INNOVATIVE STATE WETLANDS PROGRAMS

State	Legislative Authority and Regulatory Agency	Activities Regulated	Buffer Zones	Typical Mitigation Ratios	Mitigation Banking Allowed?	Assumption of Federal 404 Program?
Florida	Warren S. Henderson Wetlands Protection Act (1984); Department of Environmental Regulation	Dredge and fill	None	2.5:1–4:1[1]	Yes	No
New Jersey	Freshwater Wetlands Protection Act (1987) and the Wetlands Act (1970); Department of Environmental Protection	Removal, fill, dredge, alteration	25–150 feet (for freshwater wetlands)	1:1–7:1[2]	Yes	Working on it
Michigan	Goemaere-Anderson Wetlands Protection Act (1979); Department of Natural Resources	Removal, fill, dredge, drain	None	1:1	No	Yes
California	McAteer-Petris Act (1969) and the California Coastal Act (1976); California Coastal Commission and the San Francisco Bay Conservation and Development Commission	BCDC—chiefly beach access and removal and fill; CCC—very broad range of activities	100 feet required by CCC, none by BCDC	1:1	Yes	No
Oregon	Removal-Fill Law (1985); Division of State Lands	Removal and fill	None	1:1–6:1[3]	Yes	Decided against
Massachusetts	The Wetlands Protection Act (1972); Department of Environmental Protection	Removal, fill, dredge, alteration	Up to 100 feet	1:1	No	No

1. 2.5:1 for created wetlands, higher (4:1 and up) for enhanced wetlands.
2. 1:1 minimum, 7:1 for enhancement.
3. 1:1 for nontidal wetlands, up to 6:1 for tidal.

Like California, Michigan, and about 10 other states, Massachusetts has enacted an Environmental Protection Act. If a state permit is required and if a project exceeds one of the state's regulatory thresholds (for example, if more than 10,000 cubic yards of material is to be dredged or disposed of in a wetland), a review is triggered by the Massachusetts Environmental Protection Act. This review delays a project yet further.

Mitigation

Proposals to develop in wetlands or buffer areas must clearly demonstrate that no viable alternative to filling or altering the wetland exists. The state assigns a higher value to bordering vegetated wetlands than it does to other resource areas such as floodplains. Bordering vegetated wetlands are presumed significant for six of the seven statutory interests and, according to DEP regulations, "the complex natural functioning of these wetlands cannot be replicated, and no amount of engineering will enable such areas to be filled or substantially altered without seriously impairing the statutory interests they serve."[26] Given DEP's position, applicants will be hard-pressed to get a permit to alter such wetlands. Filling can occur in floodplains, however, if an applicant provides adequate compensatory flood storage, since this is the primary state-protected interest in floodplains and can be satisfactorily reproduced. The more statutory interests that a resource area provides, the harder it is to develop in that area.

Applicants altering up to 5,000 square feet of bordering vegetated wetland must create a wetland of equivalent size. Across the state, over 1,000 replication projects have been undertaken. The regulations decree seven conditions or standards for replacing wetlands, including the following: at least 75 percent of the surface area of the replacement wetland must be reestablished with indigenous wetlands plants within two growing seasons; and the replacement wetland must be located within the same general area as the lost wetland.

For those rare projects that receive authorization to alter over 5,000 square feet of bordering vegetated wetlands, the same replication requirements and standards apply.

Following completion of a project, the applicant may request a certificate of compliance, which certifies that the project was completed in compliance with the permit.

Monitoring and Enforcement

An order of conditions usually states that a replicated wetland must be fully established within three years. Most local conservation commissions do not, however, have the resources for regular monitoring, and when they do monitor, they do so usually only for two years.

Both DEP and a conservation commission may take enforcement action against those who violate the state's wetlands laws. Common violations include: failure to obey a particular condition or time restriction of the permit; failure to complete the work that was agreed upon in the permit; and failure to obtain a permit before altering a regulated wetland. DEP, like EPA and the Corps, can issue administrative penalties of up to $25,000 per day of violation. The Wetlands Protection Act also allows the police or a group of 10 citizens to bring a criminal action in the local district court. Fines may reach as high as $25,000 or up to six months imprisonment for each day of violation. Violators may also be forced to undo the damage they have done. One violator was sentenced to 30 days in jail, but the sentence was suspended after he agreed to restore the site to its original condition. From 1987 to July 1988, DEP issued 20 fines totaling about $500,000. The largest fine was $140,000 and the smallest was $200. Most fines are appealed. According to a DEP official, fewer fines were issued in 1988 than in previous years because the state received a record number of permit applications and consequently spent most of its time processing permits and very little on monitoring and enforcement.

In addition to the Wetlands Protection Act, Massachusetts has also enacted a Coastal Restriction Act and an Inland Restriction Act. The goal of these acts is permanently to protect the state's most ecologically significant wetlands in advance of development proposals by permanently restricting certain activities, such as dredging and filling, in such wetlands. Again, certain exemptions apply, such as outdoor recreation, installation of floats or piers, and shellfish harvesting. Under the acts, the state maps qualifying wetlands on a town-by-town basis and, following a public hearing, places permanent restrictions on certain activities in designated wetlands.

These restrictions are recorded on the deeds covering designated land in order to give notice to future property owners.

Since the Coastal Restriction Act was enacted in 1965 and the Inland Restriction Act in 1968, approximately 46,000 acres of coastal wetlands and 7,700 acres of inland wetlands have been protected. Most of the towns that participate in the restriction program are in eastern Massachusetts.

Massachusetts issues no state general permits, nor does it have mitigation banking.

Notes

1. From a personal conversation on August 30, 1989, with Stan Geiger of Scientific Resources, Inc., Lake Oswego, Oregon.

2. 53 FR 20772.

3. See The Conservation Foundation, *Protecting America's Wetlands: An Action Agenda*, the final report of The National Wetlands Policy Forum (Washington, D.C.: The Conservation Foundation, 1988), p. 23.

4. Demming C. Cowles, et al., "State Wetland Protection Programs: Status & Recommendations," a report prepared for the U.S. Environmental Protection Agency, Washington, D.C., December 1986, pp. 10, 19.

5. For further information see Cowles, "State Wetland Protection Programs," p. 26.

6. The coastal zone, as defined in the CZMA of 1972, consists of "coastal waters and adjacent shorelands strongly influenced by each other and in proximity to the shorelines of the coastal states." The zone extends inland from the shoreline and includes a variety of interdependent ecological systems, such as wetlands, lakes, streams, beaches, and dunes, "the uses of which have a direct and significant impact on coastal waters."

7. FLA Stat. 17–12.015 (1)(b).

8. Personal conversation on August 28, 1989, with William Carey of Land Sciences Corporation, Fort Lauderdale, Florida.

9. Florida Department of Environmental Regulation, "Report to the Legislature on Permitted Wetlands Projects, October 1, 1987–September 30, 1988," February 1989.

10. N.J.S.A. 13:B-27.

11. 20 N.J.R. 1264, 6/6/88.

12. See William Travis, "A Comparison of California's Coastal Programs," *Coastal Zone 87: Proceedings of the 5th Symposium on Coastal and Ocean Management* (New York: American Society of Civil Engineers, 1987), p. 2911.

13. Personal conversation on July 7, 1989, with Robert Batha, environmental planner, San Francisco Bay Conservation and Development Commission, San Francisco, California.

14. Travis, "A Comparison of California's Coastal Programs," p. 2913.

15. According to a staff report by the San Francisco Bay Conservation and Development Commission, "Diked Historic Baylands of San Francisco Bay," 1982.

16. See the California Coastal Commission's, "Statewide Interpretive Guidelines, December 16, 1981," p. 52. (Available from California Coastal Commission office in San Francisco.)

17. California Coastal Commission, "Statewide Interpretive Guidelines," p. 45.

18. San Francisco Bay Conservation and Development Commission, "Mitigation: An Analysis of Tideland Restoration Projects in San Francisco Bay," a staff report dated March 1988, p. 55.

19. From the "1988 Annual Report" of the San Francisco Bay Conservation and Development Commission, issued January 1, 1989, p. 10.
20. See Travis, "A Comparison of California's Coastal Programs," p. 2913.
21. As reported in a staff report of the Oregon Division of State Lands, "State Assumption of the Federal 404 Permit Process," December 1988, p. 2.
22. O.R.S. 541.625.
23. Act 203, P.A. of 1979.
24. The act in question is the Michigan Environmental Protection Act (1970, P.A. 127).
25. R 281.923, 3(3).
26. Massachusetts Department of Environmental Quality Engineering (renamed Department of Environmental Protection), "Wetlands Protection Act Regulations" (M.G.L. Chapter 131, Section 40), p. 3.

CHAPTER 4

MITIGATION STRATEGIES

Developers have always practiced two kinds of mitigation: the kind they do before they buy a site and the kind they do afterwards. They often avoided sites with potential environmental liabilities such as floodplains or steep slopes simply by not buying the property. Seldom, however, did developers reject a site because it contained wetlands; the only constraint was whether or not the site could be filled and developed economically. Today, though, the presence of wetlands is enough to frighten away even the most sophisticated developers. In response to increasingly stringent wetlands protection laws, developers have become more cautious in their site selection. Wary developers frequently decide against buying, or often at least reduce their offer for, property containing wetlands. One developer in North Carolina stated that if he finds wetlands on his site he will "sell the property and go find some highlands."

This chapter, however, chiefly concerns issues faced by developers who have bought property containing wetlands and who are intent upon developing it. By focusing on four types of mitigation strategies, and providing two case studies of each type, this chapter offers very practical solutions to a variety of very real problems.

MITIGATION CHOICES

Wetlands come in all shapes and sizes. Some are shaped like amoebas and refuse to fit neatly into site plans, others are geometric. Some occupy the center of a site and limit development options, others lie along the edge and may present only a minor constraint or inconvenience. The choice of mitigation strategy depends largely on the type of project, the size of the property, the size, condition, and position of the wetlands, and the regulatory environment. No single strategy will work for every site. Mitigation for a small residential development in tidal wetlands, for example, will differ considerably from that for a large office/retail complex in a freshwater wetland. Furthermore, mitigation on the West Coast will differ from that on the East, primarily because of the differences in wetlands types, but also because of historical differences in the types of disturbances to which wetlands in the two areas were subjected. Wetlands along the Pacific Coast were often diked for agriculture, while those along the Atlantic Coast were typically altered by urban development.

Generally, the bigger the site, the greater the mitigation flexibility: large sites provide room to position buildings around wetlands or to create wetlands on site, whereas small sites generally do not. Also, compared to freshwater wetlands, the relative simplicity and predictability of tidal wetlands make them better subjects for either creation or restoration; in other words, tidal wetlands offer more mitigation options.

Literally hundreds of developments have involved mitigating impacts on wetlands. In some cases, the mitigation has been very successful, while others have

> *"Any wetland that you intend to fill had better be very valuable [to your project] because you will have to spend $50,000 an acre building a new one somewhere else."*
>
> Don Tilton, Johnson, Johnson & Roy

been dismal failures. But success, like happiness, is very subjective. If success means only that wetlands plants were growing in the newly created or restored wetland or that permit conditions were met, then most mitigation efforts can be considered successful. However, if it means that all the functions and values of a filled wetland were completely restored or replaced elsewhere, then there are far fewer successes. But those criteria inject even more subjectivity: what are the values of wetlands and how can they be measured? While few observers will agree on the same criteria for success, most would agree that it is nearly impossible to replicate completely all wetlands values and functions. Some projects, however, have come impressively close—as the projects shown in this chapter demonstrate.

In general, mitigation strategies can be divided into four categories, which closely follow CEQ's definition of mitigation: avoidance/minimization, restoration, enhancement, and creation. In this chapter, each strategy is first discussed briefly and then illustrated by two examples. Each example includes a description of the project, its location, the permit process, the mitigation, and a few of the lessons learned by the developer. The examples are in various stages of completion. Some were completed several years ago, others are less than a year old, and a few are still under construction. Whatever stage the projects have reached, however, something can be learned from each one.

AVOIDANCE/MINIMIZATION

"If you want to get on base, hit the ball where they ain't."
Baseball cliché

Although it may stretch the meaning of "mitigation," the best mitigation strategy for developers is to avoid building in wetlands in the first place by placing all buildings, roads, and parking lots in upland areas and keeping the wetlands open. The benefits of this simple strategy are compelling. Numerous developers have spared themselves considerable time, cost, and aggravation by building around, rather than in, wetlands. Furthermore, preserving wetlands often adds considerable value to a project; wetlands preserved in their natural setting frequently turn out to be one of the most attractive features of a project.

Probably the most convincing argument for preserving a wetland is that it is considerably cheaper than creating a new one. Wetlands generally cost anywhere from $20,000 to $75,000 an acre to build; and occasionally, developers must build a replacement wetland on what would otherwise be a choice piece of real estate. In Michigan, for example, a developer had earmarked a seven-acre site adjacent to a marina he was developing for condominiums; but the seven-acre site was the only available location for the on-site mitigation required by the state, and so the developer had to create a wetland instead of the condominuims.

One popular method of mitigation is to cluster buildings in one area while keeping the rest of the site open. If permitted by local zoning, this approach can cut down the magnitude of the wetlands fill, or avoid the wetland entirely, and save on infrastructure costs as well. This is not a new concept. In Lincoln County, Massachusetts, every subdivision approved by the planning board during one 10-year period in the 1970s and 1980s was a clustered development. Although increasing in popularity, clustering is still prohibited by some jurisdictions and allowed by others only through a special exemption. Generally, urbanized counties, especially those that allow Planned Unit Developments (PUDs), will accommodate clustering, whereas their rural counterparts will not. Subdivision regulations in Fairfax County, Virginia, for example, allow clustering for subdivisions. The county requires that, depending on the type of zone, between 15 and 50 percent of the clustered subdivision be designated as open space. According to a Fairfax County zoning official, over 80 percent of recent developments of subdivisions in the county were clusters.[1]

Harford County, Maryland, likewise requires developers to set aside a minimum percentage of their residential clusters as open space. For example, 10 percent of a site must be left open in the county's R-1 and R-2 residential zoning districts, 15 percent in R-3, and 20 percent in R-4. Only 60 percent of the open space within a project can be comprised of passive uses and the rest must be set aside for active uses, such as ballfields and tot lots. Wetlands may provide part of the passive use. In addition, the county's policy regarding development in its designated natural resource districts offers incentives for developers to cluster buildings and protect sensitive environments. Clusters make up the bulk of recent residential developments in Harford County (see The Villages of Thomas Run case study on pp. 72–74).

Although building adjacent to, rather than in, wetlands will reduce the adverse environmental impacts as well as regulatory burdens, it will not guarantee the long-term survival of the spared wetland. A wetland is not an isolated, independently functioning ecosystem; changes in its surrounding environment greatly influence the character and well-being of a wetland. Sediment and polluted runoff from nearby development can destroy a healthy wetland. For this reason, many states require developers to establish a buffer zone of 100 feet or so between the development and the wetland.

Damage to a wetland and its inhabitants can be reduced by utilizing certain "wetland-friendly" con-

THE CLUSTER CONCEPT

A medium-sized residential project in Atkinson, New Hampshire, illustrates the cluster concept. In this project, the developer utilized a simple but effective strategy to deal with the numerous wetlands scattered over the 111-acre site: build where the wetlands are not. Some filling was necessary for roads, but the amount to be filled was kept below 10 acres in order to qualify for a general permit from the Corps.

The project, called The Commons at Atkinson, contains 98 townhouses clustered on the upland portions of the site and features about 90 acres of open space, 78 of which comprise the "town greens." The designers have dealt with the Corps before, so they know what has to be done to get a project approved. As Gordon Leedy from Matarazzo Design remarked: "Once you go through the 404 process a few times and get banged up by the regulators, you learn what you can and cannot do. Time is money so you learn real fast."

SITE PLAN OF THE COMMONS AT ATKINSON

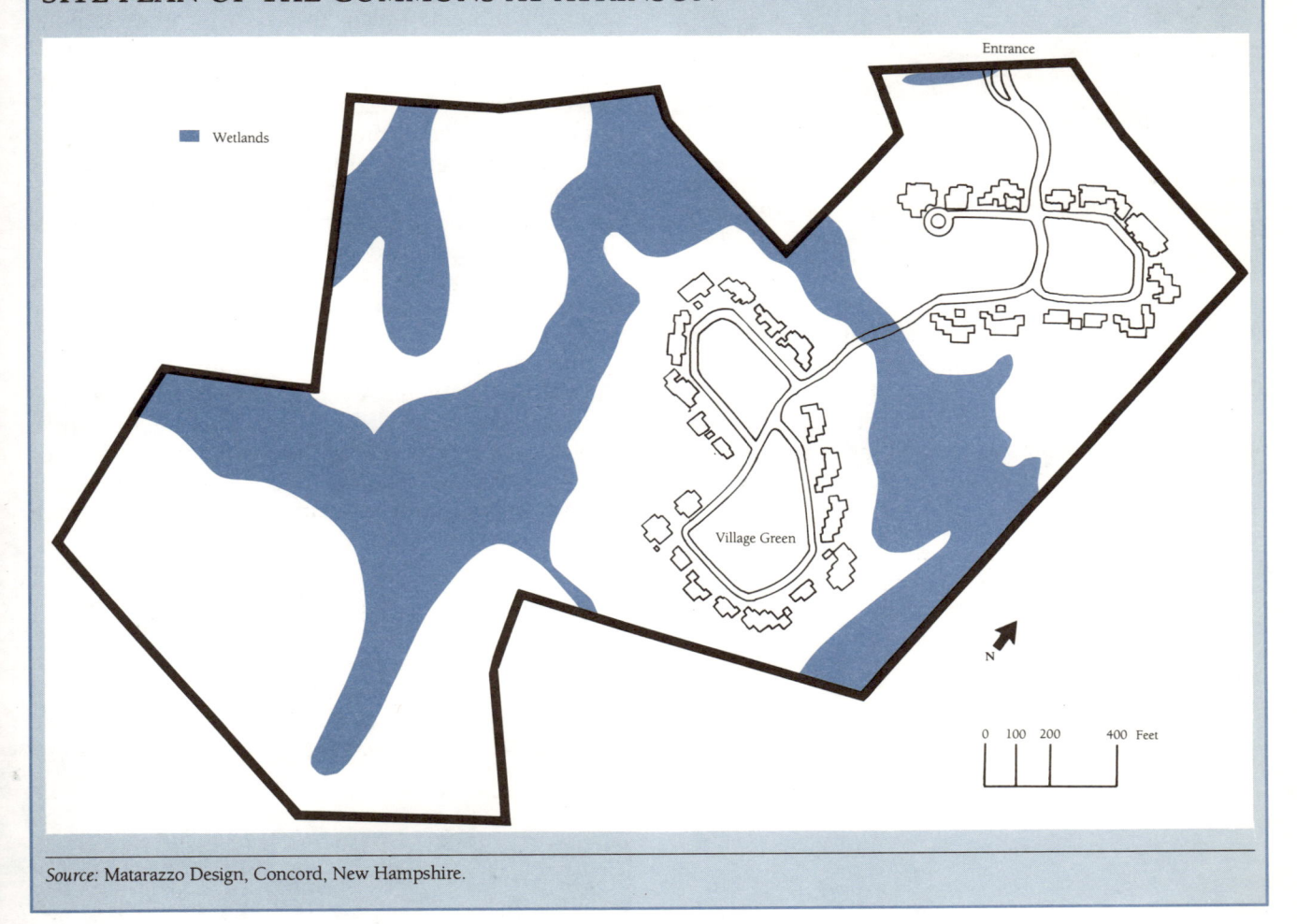

Source: Matarazzo Design, Concord, New Hampshire.

struction techniques and materials, and by carefully scheduling the timing of construction so that it does not interfere with, for example, bird mating and nesting seasons. One developer imported special low-ground-pressure grading and support equipment from England to minimize the damage development would otherwise cause to a marsh. Others developers have built temporary, low-impact roads and work stations out of logs or steel construction plates to reduce the damage caused by heavy construction equipment.

In a few instances, the Corps has actually built temporary dikes to dry out a wetland so that heavy equipment could be moved in without overly compacting the wetlands soils. After the heavy work was completed, the

Heavy earth-moving equipment can severely compress soft, muddy soils and irreparably change the hydrology of a wetland. Moving cranes along temporary log roads, as shown here, distributes the weight over a greater area and significantly reduces the degree of soil compression.

equipment was removed, seedlings were quickly planted in the still moist soil, and the dike was breached, allowing the water to soak the site once again.

The following two examples illustrate how developers were able to build on sites containing wetlands while avoiding or at least substantially minimizing any wetlands fill.

THE VILLAGES OF THOMAS RUN

Harford County, Maryland's zoning ordinance is unusual in that it allows developers to "bump up" to the next highest residential zoning classification if more than 30 percent of a site is designated as a natural resource district. The purpose of such districts is to preserve special environmental features such as wetlands, streams, rivers, and steep slopes. Within natural resource districts, nontidal wetlands "shall not be disturbed by development." In addition, developers must maintain a 75-foot buffer between development and wetlands.

A local developer, Oak Investment Company, Inc., originally submitted plans to Harford County to build 430 single-family homes on individual lots on a 151-acre site characterized by gently rolling terrain and dissected by two small streams, along which lie over 20 acres of bottomland forested wetlands and wet meadows. Over 30 percent of the site was designated as a natural resource district. The proposed development would have entailed extensive wetlands filling, five stream crossings, and would have eliminated most of the mature woods on the site. Citing the project's likely adverse impacts on wetlands, the county rejected the proposal outright.

After being rebuffed by the county, the developer hired Morris & Ritchie Associates, Inc., a local planning and engineering consultant, to develop a site plan which would accommodate the site's constraints and take advantage of the special natural resource district provisions in the county's zoning ordinance. The revised proposal

The natural resource district provisions of Harford County's zoning ordinance provide incentives for developers to preserve wetlands.

called for clustering 530 townhouses on upland portions of the site and leaving the wetlands virtually untouched.

Because more than 30 percent of the Thomas Run site lies within a designated natural resource district, Harford County's zoning ordinance allowed the site to be developed with lots, setback requirements, and housing types allowed in the next higher zoning classification, but at no greater overall density than allowed by the original zoning. The Thomas Run site was in an R-2 zone, but its natural resource lands bumped it up to an R-3. Thus, instead of building 24-foot-wide townhouses in groups of four or less, the developer could build 20-foot-wide townhouses in groups of eight. By clustering the units on the upland portions of the site, the developer was able to preserve nearly half the site as passive open space, reduce stream crossings from five to three, and greatly reduce the costs of grading, roads, and utilities. In addition, because no wetlands were filled except near the stream crossings, the developer qualified for a nationwide permit—a generic permit from the Corps that is easier to obtain than an individual permit.

Following county approval of the revised plan, the development was sold to Rachuba Enterprises, Inc., and by the end of 1989 nearly half the planned units at Thomas Run had been built.

By clustering the townhouses, the developer was able to preserve nearly half the site, including over 20 acres of wetlands, as passive open space.

FIGURE 4.1
INITIAL AND APPROVED SITE PLANS FOR THE VILLAGES OF THOMAS RUN

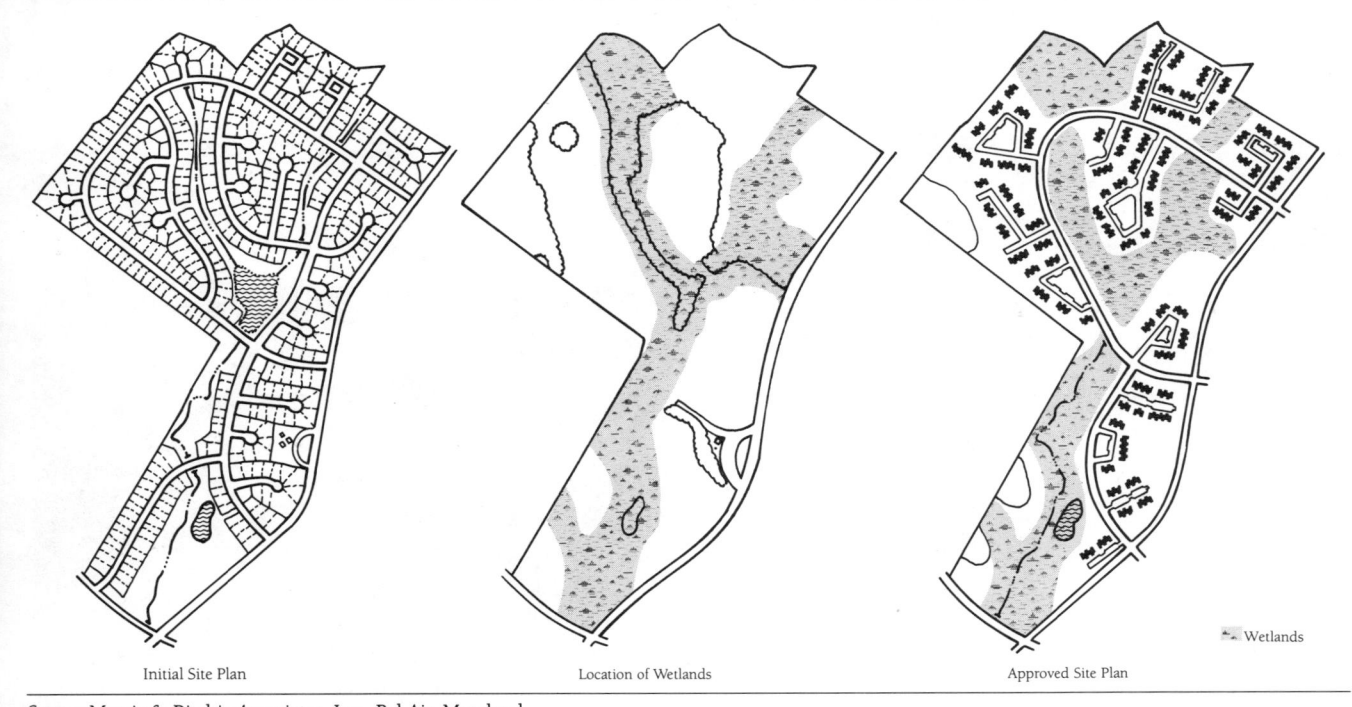

Initial Site Plan   Location of Wetlands   Approved Site Plan

Source: Morris & Ritchie Associates, Inc., Bel Air, Maryland.

Harford County's provisions allowing developers to "bump up" to the next zoning classification offer strong incentives for developers to preserve sensitive environmental features. According to Tom O'Laughlin of Morris & Ritchie Associates, "Developers now seek land with natural features such as wetlands and steep slopes because it allows them to build a greater mix of housing types at greater net densities than would otherwise be permitted."

Lessons Learned

- Do not automatically shy away from property containing wetlands. Because of the flexibility in Harford County's zoning for natural resource districts, developers can increase the number of units they can build while preserving sensitive environmental features.
- Through careful site planning, developers can work with most environmental constraints.

PROJECT DATA

Developer
Rachuba Enterprises, Inc.
Towson, Maryland

Builders
Ryland Homes, Inc.
Lutherville, Maryland

Pulte Home Corp.
Silver Spring, Maryland

Site Planner and Engineer
Morris & Ritchie Associates, Inc.
Bel Air, Maryland

Wetlands Consultant
EA Engineering, Science, & Technology
Sparks, Maryland

Geotechnical Consultant
Geo-Technical Associates
Bel Air, Maryland

Montecito Apartment Village

At first glance, the biggest drawback to this 40-acre parcel located in Tampa, Florida, was the 10-acre wetland located smack in its center. But Embrey Investments, Inc., of San Antonio, Texas, was able to turn the wetland from an apparent liability into a unique asset. Today, the 384-unit apartment village features an attractive wetland with beautiful cypress trees as its centerpiece. Instead of filling the wetland, the developers used it to their advantage.

Architects Fusch, Serald & Partners of Dallas designed the 45 two-story apartment buildings. The units are built in three clusters or pods that sit on the upland portions of the site and avoid the wetland entirely. The three clusters surround the wetland area and afford 70 percent of the units views of the greenery and other amenities. The wetland contains a cypress swamp and a three-acre lake.

Each building contains between eight and 10 units, with rents ranging from $390 to $670 per month. The design and location of Montecito appeal to young professionals who commute to downtown Tampa. The village features an exercise facility, clubhouse, pool, jacuzzi, tennis courts, and, around the central wetland, a two-mile jogging trail. The gross density of the project is 10 units per acre.

Permitting

Embrey Investments purchased the site in 1987 and began construction that year. No permits were needed from the Corps because the project necessitated virtually

Preserving wetlands often adds value to a project. Many apartments at Montecito enjoy views of the lake.

Clustered development can preserve valuable open space, such as this 10-acre wetland. By not building in the wetland, the developer avoided the rigorous Section 404 permit process.

no wetlands fills. Unavoidably, however, a small sliver of wetlands within the "waters of the state" was filled. Consequently, the developer had to obtain a permit from Florida's Department of Environmental Regulation and the Hillsboro County environmental protection office. Hillsboro County's wetlands laws require that all artificial forested wetlands be monitored for five years after their creation, and that during that period at least 85 percent of the wetlands plants survive. If more than 15 percent of the plants die each year, the developer is required to replace them with new plants. The county also requires a 30-foot setback from all wetlands in which neither construction nor impervious surfaces are allowed. The jogging/walking trail that passes through the setback area is made of mulch, and a section of the village's parking lot that penetrates the buffer zone was paved with porous concrete.

Mitigation

Embrey Investments avoided the rigors of the Corps permit process by developing around, rather than in, the wetlands. Still, some mitigation was required. The developer had to pave a small area of wetlands to provide access to parking facilities. In exchange, an upland area of about one-half of an acre was dug out, graded, and planted with trees. The developer removed up to one foot from the surface of the mitigation site and replaced it with about six to 12 inches of organic soil taken from the filled wetland. The area was then planted

In addition to the tennis courts and two-mile walking/jogging trail, the project also features a jacuzzi and swimming pool.

with over 100 trees, primarily pond cypresses and an assortment of red maples, black gums, sweet bays, and swamp bays. The developer also preserved 150 or so mature oak trees on site. The trees were temporarily transplanted to the wetlands area for safekeeping during construction, and then replanted throughout the site when construction was complete. An official of the county Development Review Department, which generally frowns on any proposals to remove mature trees from wetlands, commented that Embrey has done a "good job of saving existing trees." Embrey also had to install a pipe in the area beneath the road where the

FIGURE 4.2
SITE PLAN OF MONTECITO VILLAGE

Source: Embrey Investments, Inc., San Antonio, Texas.

wetland was filled so that small animals could cross the road unharmed.

Although some minor problems have surfaced with the created wetland (the grading, for example, was not done properly so too much water lies in pools around the trees), both the state and county seem pleased with the project. The DER permit was issued in July 1987.

Lessons Learned

- Wetlands can actually increase the value of a project.
- Save the large specimen trees where possible to prevent the project from looking barren during its early years.

PROJECT DATA

Developer
Embrey Investments, Inc.
San Antonio, Texas

Architect
Fusch, Serald & Partners
Dallas, Texas

Environmental Consultant
Biological Research Associates, Inc.
Tampa, Florida

RESTORATION

"And restore the tone of languid Nature."
William Cowper

In most cities and towns, the best sites—the dry, relatively flat, stable sites—were developed first, and those that were more difficult or expensive to develop—wetlands, steep slopes, flood-prone areas—were left for others to pick over later. Many of the leftover sites were diked, dumped on, or filled years ago. Suffocating from fill, deteriorating from pollution and debris, these abused wetlands retain only a modicum of their full range of ecological values. These wetlands can, however, be returned to ecologically productive uses by cutting off pollution discharges, removing fill, regrading the site, and introducing native wetlands plants. Unexpectedly but increasingly, these wetlands are becoming attractive targets for development. Even degraded wetlands, however, are protected by state and federal laws, although usually not as vigorously as pristine wetlands. In exchange for filling a wetland on site or nearby, developers have restored a number of these long-neglected wetlands to thriving, self-sustaining ecosystems.

Some states prefer that developers restore rather than create a wetland as compensation for filling another wetland, since restoration, these states believe, involves less risk. Conversely, others now protect their altered or degraded wetlands and will no longer accept restoration as mitigation; these states require that developers create a new wetland out of upland and thereby produce a net gain in wetlands. The logic behind this second approach is that restoring a degraded wetland in exchange for filling a perfectly good one leads to a net loss of wetlands acreage and values; a wetland, however damaged, is still a wetland. By contrast, no net loss of wetlands values or acreage will occur if a new wetland is created out of upland, assuming that the created wetland exhibits the same values as the one that was filled.

In the Hackensack Meadowlands of New Jersey, not far from Manhattan, a 7,500-acre abused and neglected wetland is one of the few, and perhaps the only, remaining large undeveloped tracts of land in the area. The area is under intense development pressure with property values running as high as $500,000 per acre. One company, Hartz Mountain, has developed nearly 3 million square feet of commercial and industrial space in the Meadowlands and is planning to add over 3,500 houses in the near future. The project is possible only because of the extensive mitigation involved. In exchange for filling 128 acres of wetlands, Hartz is required to restore and preserve 151 acres of severely polluted, unproductive wetlands. Thus far, Hartz has created a 63-acre salt marsh out of polluted muck that

PORTRAITS OF TWO MARSH RESTORATIONS IN THE SAN FRANCISCO BAY

Warm Springs Marsh
Fremont, California

Located at the southeast corner of the San Francisco Bay, the 250-acre King and Lyons Business Park overlooks a budding 260-acre restored tidal marsh called Warm Springs Marsh. In the 1930s, the marsh, like so many other tidal marshes along the San Francisco Bay, was diked off from the Bay and converted to cattle pasture. In its previous natural state, the marsh was probably a salt marsh dominated by pickleweed. Behind the dikes lay "seasonal wetlands"—shallow depressions that fill with water during the fall and winter rains and dry out each spring and summer. Supporting a unique collection of plants and animals that can tolerate the annual transformation from desert to freshwater pond, these short-lived wetlands were scattered across the site. According to the Corps, the site contained pockets of seasonal wetlands in upland areas and pockets of upland in wetlands areas. After the Corps issued a Section 404 permit in 1984, a consultant, Philip Williams and Associates, devised a restoration strategy that was designed to consolidate the wetlands in one area and the

Salt marshes generally take longer to develop in California than on the East Coast. During the dry season, the amount of freshwater flowing from the mountains to the ocean is significantly reduced, resulting in a greater salinity of the coastal Pacific waters. The greater salinity slows the rate of plant growth. Three years after the dike was breached, less than half of Warm Springs Marsh was covered with plants.

SITE PLAN OF WARM SPRINGS MARSH

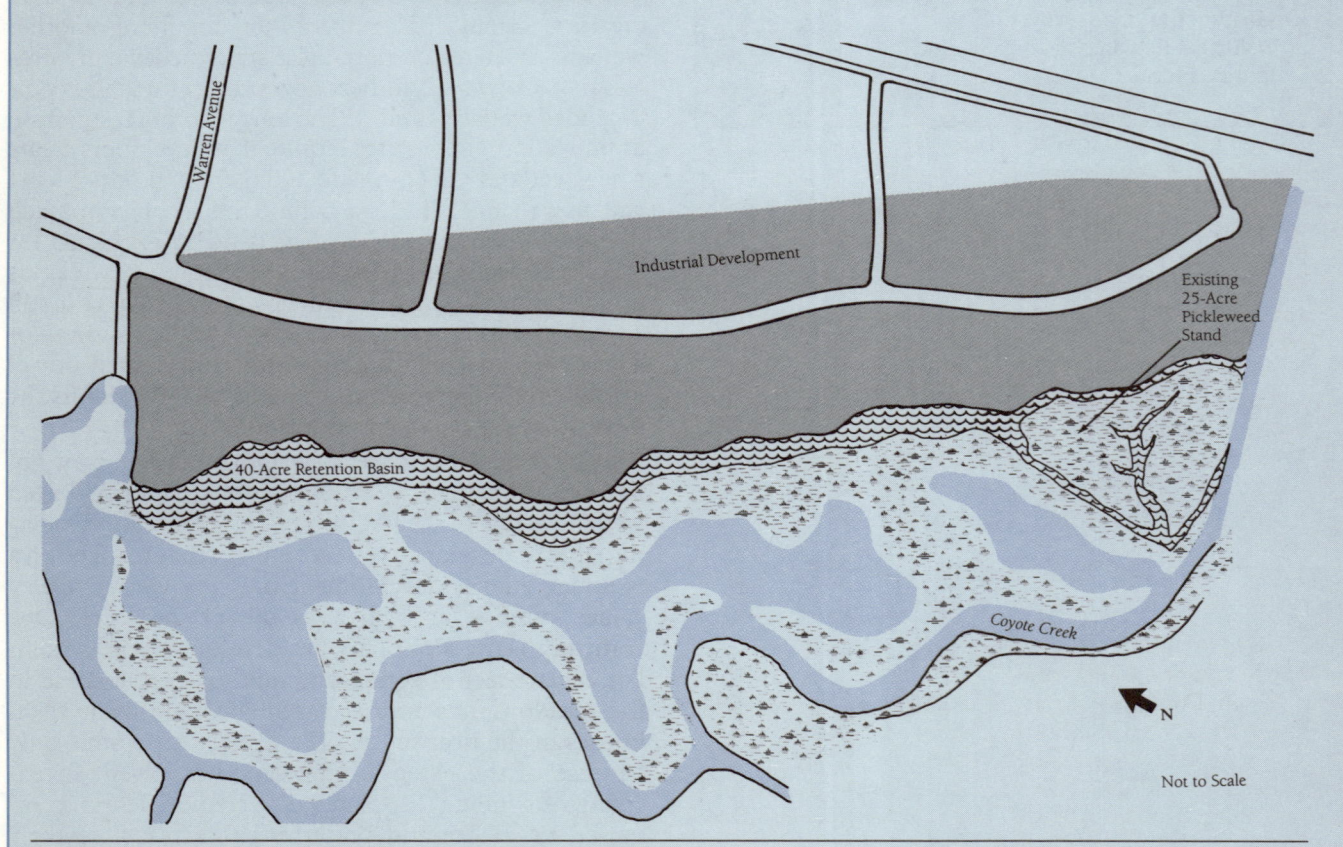

Source: Thompson + Wright, Architects and Planners, San Francisco, California.

upland in another. In exchange for filling over 200 acres of seasonal wetlands, the developer was required by the Corps to restore tidal action to the diked marsh and to leave intact a 25-acre stand of pickleweed.

The goal of the restoration was to create the right hydrological conditions for different plants to prosper and then to let Nature do the rest. In order to construct the building pads for the business park, which was designed to contain light manufacturing as well as R&D facilities, over 2 million cubic yards of sand and silt were excavated from the marsh restoration site as parts of the site were lowered to about 10 feet below sea level. Excavation occurred at three different locations to create three distinct basins, each checkered with holes and surrounded by a 100-foot-wide shelf that was constructed to support plants like alkali bulrush. After excavation and grading were completed, in 1986 the dike was breached in three places and the Bay waters quickly swept over and reclaimed the long dormant marsh.

As the accompanying figure shows, since the business park is located within the 100-year flood area, the developer constructed a flood control levee inland of the restoration area. Between the levee and the industrial park lie two separate managed areas: the untouched stand of pickleweed and a 40-acre linear stormwater retention basin.

The southeast portion of the Bay experiences very high sedimentation rates (as high as one foot per year), probably the highest in the Bay. According to an official at Philip Williams and Associates, as sedimentation proceeds, the restored marsh should evolve relatively quickly through a series of wetlands types: from mudflats to salt marsh dominated by cordgrass, to a brackish marsh dominated by alkali bulrush marsh, and finally to a fully developed pickleweed marsh. Within two years, pickleweed had begun to colonize the basin shelves. A botanist who inspected the site, however, noted that colonization is proceeding slowly, with large areas of the shelves still bare.[1]

With over 200 acres of marsh restored, this project qualifies as one of the largest restoration efforts ever undertaken and one of the last large projects permitted in the San Francisco Bay. As far as the Corps is concerned, it is one of the most successful projects as well. When the project was first conceived in the early 1980s, seasonal wetlands received scant protection from the Corps. Not so anymore. In today's regulatory climate, the project probably would not be permitted.

Muzzi Marsh
Marin County, California

When Dominic Muzzi diked 225 acres of tidal salt marsh along the San Francisco Bay in 1959, he would surely have been surprised to know that, 25 years later, someone would spend over $2 million to buy the site and turn it back into a wetland. But when the Golden Gate Bridge, Highway, and Transportation District decided to build a new ferry terminal on the Corte Madera Creek, it

Thirteen years ago, the dikes along Muzzi Marsh were breached. Today, this thriving salt marsh supports a variety of plants, fish, and shorebirds.

needed a site both to mitigate for dredging over 30 acres of mudflats and salt marsh for the access channel from the San Francisco Bay and to dispose of the dredge spoils. The nearby and still vacant Muzzi site was an ideal choice. Of the 200 acres purchased by the District in 1976, 70 were used for dredge disposal and 130 for marsh restoration. Some of the dredge spoils were used to shape the marsh prior to its restoration.

In 1976, the dike was breached in four places and the Bay waters once again washed over the site with the ebb and flow of the tide. Although no planting or seeding was done, within a year pickleweed began invading and colonizing the marsh followed by cordgrass a year or two later. By 1982, both pickleweed and cordgrass were well established. Natural revegetation proceeded too slowly, however, in the opinion of the California Department of Fish and Game and the City of Corte Madera, so in 1980 the District modified the site to increase tidal access and to restrict human, dog, and cat intrusions. From each of the four breaches, the district dug oversized channels to create two loops around the periphery of the marsh—like moats around a castle. One year later, cordgrass colonized the edges of the new channels and, by 1986, large stands of cordgrass were established.

According to Phyllis Faber, a botanist who has monitored the Muzzi Marsh restoration since its inception, the restoration was successful in creating quality wildlife habitat, but she believes its long-term survival is doubtful. Flocks of waterfowl and shorebirds regularly visit the marsh and two or three rare or endangered species inhabit the marsh as well. Unfortunately, due to heavy sedimentation, the channels are fast filling with sediment and the marsh seems to be evolving rapidly from a wetland to an upland habitat.[2]

Notes:
1. Personal conversation on October 2, 1989, with Phyllis Faber, a wetland biologist living in San Francisco, California.
2. Ibid.

The Hackensack Meadowlands lie in one of the most heavily developed corridors in the nation.

was a haven for the upstart common reed (considered a weed) but not much else. Although it is still too early to tell, by many accounts the mitigation seems successful. Birds have responded quickly to the restoration: about seven times as many birds and twice as many species populate the restored site compared to an adjacent, unrestored site. The restoration has not, however, been cheap. In fact, it is one of the most expensive private wetlands mitigation efforts ever undertaken; Hartz has spent over $5 million already, or over $75,000 per acre.

Restoration is often as simple as removing the dike, drain, or fill that altered the wetland in the first place. Many diked wetlands along the California coast were restored merely by breaching a dike and exposing the wetland to tidal action. Other wetlands were restored by creating the right environmental conditions in which desirable wetlands plants could prosper.

Some restorations are more complicated, however, and require an elaborate water-control system to provide an artificial environment conducive to wetlands plants. This process may be necessary in an enormous proposed wetlands restoration project on the Pacific Coast near Los Angeles that rivals Hartz Mountain's in scale and expense. The 960-acre site, owned by Howard Hughes Properties, an affiliate of Summa Corporation, contains 216 acres of degraded fresh and saltwater marshes called the Ballona Wetlands. Seepage from oil and gas production, runoff from roads, and chemicals from agricultural activities have damaged these wetlands. Howard Hughes Properties has proposed to restore and preserve the 216 acres as part of a mixed-use development called Playa Vista. The proposed mitigation is estimated to cost $8 million and will take about five to 10 years to complete. If the project is approved, Hughes will deed the 216 acres to the National Audubon Society, which will manage the restoration effort as well as a proposed on-site interpretive center—an integral part of the restoration. The mitigation will include a programmable, automated tide-gate control system to establish and maintain desired tidal water exchange rates. If completed, the Ballona Wetlands will become one of the largest wildlife sanctuaries along the California coast.

One of the lessons learned from the Hartz Mountain and the proposed Hughes projects is that severely manipulated and degraded wetlands can be very expensive to restore. Both Hartz and Hughes are trying to undo decades of environmental maltreatment, with no prior experience with restoration projects of this scale, no blueprints to work from, and no guarantees that the work will be successful. Nonetheless, the track record of restoration projects continues to improve.

Several recent studies illustrate why some mitigation efforts succeed while others fail.

- In a review of 14 tideland restoration projects in the San Francisco Bay Area, the Bay Conservation and Development Commission found that six of the projects were deemed successful, five were only partially successful, and three had failed. The primary reason why the three restoration projects failed was that some portion of the required work was not performed: some well-designed projects were never fully implemented. For example, in one project involving a small fill, a developer was required to create a one-quarter-acre marsh to offset the effects of placing 6,000 square feet (two-fifths of an acre) of fill in a tidal inlet. According to the BCDC report, this project failed chiefly because at least a portion of the work was never carried out. The problem lay in the permit itself, which *authorized* removal of fill but did not specifically *require* it. In addition, the permit did not set a specific date for completing the mitigation. Although a plan for removing the fill was approved in 1983, a site inspection in 1987 found little evidence that any fill had been moved.

In another example, the city of Vallejo, California, was required to return 63 acres of seasonal wetlands along the Napa River to tidal action in exchange for filling about 11 acres of wetlands and for dredging an unspecified area of marsh and mudflats in the Napa River for a marine construction facility. BCDC granted the original permit in 1974 for a small fill and extensive dredging, and authorized additional dredging and filling over the next several years. The city acquired a 63-acre mitigation site, but as of March 1988

FIGURE 4.3
SCHEMATIC ENGINEERING PLAN FOR BALLONA WETLANDS

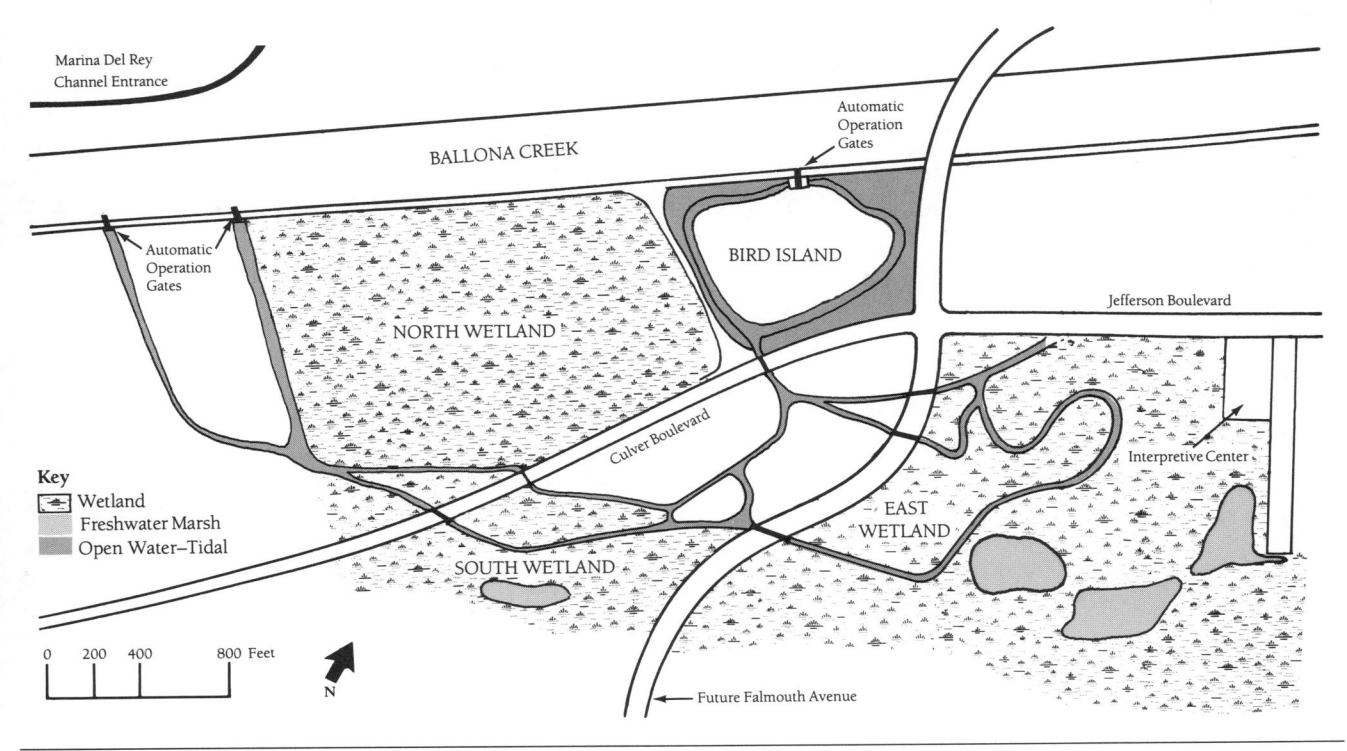

Source: National Audubon Society, "Ballona Wetland Habitat Management Plan," facing p. 24. Draft submitted to the City of Los Angeles for Review and Adoption, November 1986.

had not performed the required work. The BCDC report noted, however, that natural forces had returned much of the 63-acre site to tidal action.[2] One of the problems with this project was that the city was allowed to dredge and fill without completing the required mitigation. Perhaps BCDC should have required the city to perform the mitigation in advance of, or concurrent with, the dredging and filling.

- A study of 63 restoration projects completed in coastal California between 1954 and 1985 discovered that 65 percent of the projects were wholly successful, 25 percent were partially successful, and only 10 percent were entirely unsuccessful. Success varied significantly according to the type of organization responsible for the restoration: private sector restorations had a higher success rate (63 percent) than those of federal (50 percent) and state (40 percent) agencies. The successful projects generally benefited from accurate preconstruction elevation analyses and some post-construction monitoring, and were completed by the private sector or local agencies. Typically, the unsuccessful projects suffered from poor pre-construction engineering analysis, infrequent monitoring, and skimpy review by regulatory agencies. The study evaluated success according to two criteria: whether or not certain wetlands functions were similar between the restored and natural wetlands; and whether or not the project met its goals.[3] Increasingly, the private sector, is taking the lead in wetlands restoration and seems to be doing a better job of restoration than is government.

Although it promises to bring dead wetlands back to life, restoration it not without its drawbacks. In general, by allowing developers to fill perfectly good wetlands in exchange for restoring those of poor quality, acre-for-acre restoration leads to a net loss of wetlands because even damaged wetlands usually have some value, however slight, for wildlife, flood control, or recreation. On the other hand, some disturbed wetlands have lost their valuable qualities through years of abuse, and any restoration is therefore a plus. In any case, to compensate for the loss in values and for the possibility that not all of the restoration will be successful, regulators increasingly demand that the size of the restored area must be larger than the area being filled. Just how much larger often depends on the nature and extent of the wetlands values lost due to fill and the likelihood of success of the restoration. Regulators typically rely on fixed, but somewhat arbitrary ratios of wetlands restored to wetlands filled, since it is difficult to measure wetlands values accurately and objectively. Other regulatory authorities determine the restore-to-fill ratio on a case-by-case basis.

The following two examples, one from Florida and the other from Wisconsin, illustrate how developers have successfully restored wetlands.

BAYPORT PLAZA
TAMPA, FLORIDA

On the shores of Old Tampa Bay in Florida lies a $75 million, 46-acre office/hotel complex called Bayport Plaza. Located just a stone's throw from Tampa International Airport and only about four miles from downtown Tampa, the project boasts one of the area's most convenient and attractive business locations. It includes an 11-story, 259,000-square-foot office building, a 400-room Hyatt Regency Hotel, and over 25 acres of thriving wetlands. Before development, the site was considered an eyesore—a weed-infested, badly abused area that contained old cars, mattresses, bathtubs, and other debris that people had illegally dumped for years.

The trouble started back in the 1960s when a Memorial Highway bridge was built across a small tidal creek, called Fish Creek, that linked a tidal pond with Old Tampa Bay. A culvert was placed under the bridge to allow water from the bay to flow in and out of the pond. Gradually, however, the culvert became so obstructed with debris that the water flowing through it was reduced to a trickle. Cut off from the ebb and flow of the salt water, the brackish pond and marsh was gradually converted to a freshwater pond and marsh. Native salt-tolerant plants died and were replaced by aggressive, invading, weedy plants such as Brazilian Pepper

The 259,000-square-foot office building and 400-room Hyatt Hotel stand less than 75 feet from a mangrove swamp in Tampa Bay. The swamp contains an abundance of birds such as roseate spoonbills and snowy egrets. The birds seem oblivious to the noisy airplanes departing from and arriving at nearby Tampa Airport.

FIGURE 4.4
BAYPORT PLAZA SITE PLAN

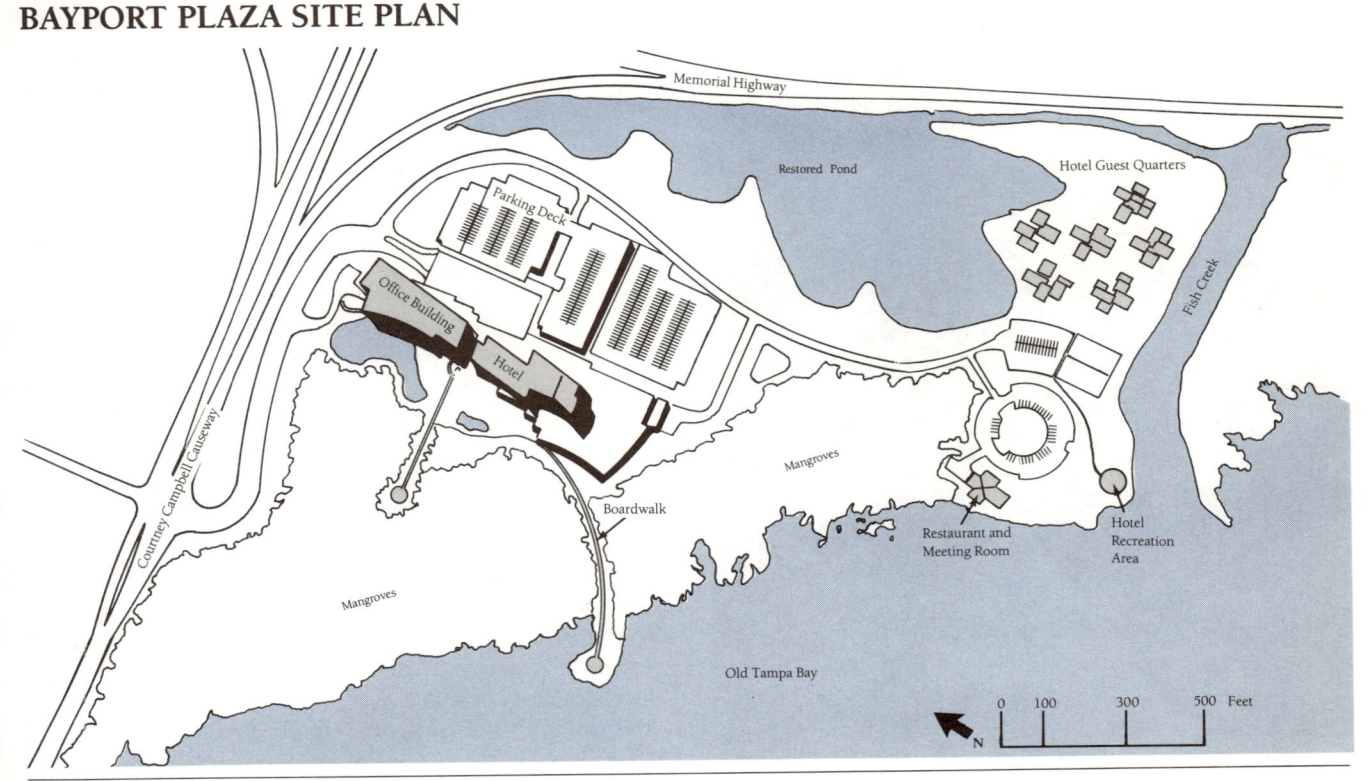

Source: The Wilson Company, Tampa, Florida.

and the ubiquitous cattail, which grow in dense stands at the expense of other plants. Fish, shrimp, crabs, and other sea creatures were doomed. And when they disappeared, so did the bevy of shorebirds that depended on them for food.

In 1984, however, The Wilson Company restored the tidal pond and its surrounding wetlands and preserved over 70 percent of the site as open space. Other developers had ignored the site because it contained wetlands and because it was a popular dumping ground. But it is also a prime commercial location: close to the airport and downtown, and right on the water.

Restoring Life to a Damaged Site

It took about two and one-half years to obtain the 27 permits that were required for project approval. The four main permitting agencies were the Tampa Port Authority, the Corps, the Florida Department of Environmental Regulation (DER), and the Hillsborough County Environmental Protection Commission. The agencies required only minor changes in the company's original plans. For example, the proposed entranceway was moved to a new location to avoid creating a traffic problem. But moving the entranceway meant that about two acres of wetlands would have to be filled. In exchange, The Wilson Company created two acres of wetlands on site. The company was to some extent fortunate that it was not developing a pristine wetland; the regulatory agencies would have been much tougher. Since the site was so badly degraded and was becoming environmentally sterile, the agencies generally looked very favorably on Wilson's development and mitigation plans.

The Wilson Company bought an option on the site in 1982 and hired an ecologist to undertake an environmental assessment. The ecologist walked the site to identify wetlands and any other potential limitations. The company wanted to know the opportunities and limitations of the site before making an offer to buy it. Of the approximately 45 acres, only 13 were considered developable. The remaining 32 acres would remain undeveloped. Satisfied that it could work with those constraints, Wilson bought the site in 1984.

The mitigation began in 1984 and was completed before the hotel was opened. The company's goal was to restore the tidal pond and adjacent wetlands to something resembling their original condition. The first thing it did was take out the trash; it hauled out over 200 truckloads. Next, it dredged Fish Creek, dug a new

An elevated boardwalk leads through the mangrove swamp to a gazebo, which sits on the site of a former radio tower.

channel connecting the creek to the pond, and uprooted and cleared dense thickets of Brazilian Pepper. In place of the peppers, the company planted native shrubs such as wax myrtle and sea myrtle, along with an assortment of marsh grasses. Returning tidal ebb and flow to the pond killed off the dense stands of cattail, which cannot withstand the higher salt concentrations of sea water. In their place sprouted native mangrove plants, which rapidly recolonized the site naturally.

The company also created two acres of salt marsh out of uplands by carefully excavating a filled marsh to the proper depth and hand-planting cordgrass sprigs throughout. (The marsh had previously been filled to provide road access to radio towers.) Over the years, mangroves colonized the edge of these filled areas. In addition, the company built a boardwalk through the marsh leading to a gazebo built on the foundations of an old radio tower. The whole mitigation process took about three years. Altogether, in exchange for filling about two acres of marsh, the company created over two acres of new marsh and restored approximately 15 acres of wetlands, including a six-acre tidal pond.

Initially, the state and local agencies were skeptical that the pond restoration would work. But when the restoration was completed, the project was judged a success. Indeed, in 1985, it received the Florida Audubon Society's Corporate Award for an environmentally sensitive development, the first time the award was given. Long-absent shorebirds quickly returned to the refurbished pond, apparently unconcerned by the presence of the office/hotel complex. According to Debora Kohne of Florida DER: "Although the site is bordered by a highway on one side and a hotel on the other, you can still find just about every kind of shorebird there. I would give the project a blue ribbon."

The Wilson Company estimates that it spent about $250,000 to evaluate the environmental conditions of the site, restore the pond, and create about two acres of salt marsh, and an additional $1 million to purchase the Fish Creek property, the area now occupied by the marsh pond, and four adjacent acres of upland. The investment was well worth it. The restoration provides an aesthetically pleasing environment that attracts tenants. The building was fully leased within three years. Like many other successful mitigation projects, The Wilson Company hired an ecologist very early in the planning process to identify and delineate wetlands, develop a restoration plan, and guide it through the environmental permit process. Today, this is standard practice in Florida.

Lessons Learned

- Do your environmental homework upfront. Before you purchase a site, find out if it contains wetlands or

After years of decay, this brackish marsh was brought back to life when The Wilson Company dug a channel reconnecting the wetland to Tampa Bay.

other potential limitations and adjust your purchase price and development plans accordingly.
- After carefully thinking through the development plans, meet with some public interest groups to solicit comments on the plans. At Bayport Plaza, The Wilson Company met with the Audubon Society, the Sierra Club, and the local Environmental Coalition. Developers must display concern for the environment, be knowledgeable about the issues, and follow through with their commitments.

PROJECT DATA

Developer
The Wilson Company
Tampa, Florida

Architect
Thompson, Ventulett, Stainback & Associates
Atlanta, Georgia

Environmental Consultant
Mangrove Systems
Lutz, Florida

LAKEVIEW CORPORATE PARK
KENOSHA COUNTY, WISCONSIN

The Des Plaines River originates in the farming country of southeastern Wisconsin and winds its way south before entering the Chicago metropolitan area. Adjacent to the river, WISPARK, a subsidiary of the Wisconsin Energy Corporation, is developing a corporate park in Kenosha County, Wisconsin. The complex, called Lakeview Corporate Park, includes a 1,200-acre business park, a 150-acre office park, and a 600-acre conservation area on the Des Plaines River floodplain. In addition, a 100-acre lake, formerly a gravel pit, lies immediately northwest of the business park.

From a business standpoint, the site is ideally located midway between Chicago and Milwaukee, one mile north of the Illinois border and only 1.5 miles from I-94, but it had one drawback: the river prevented direct

FIGURE 4.5

LOCATION MAP OF LAKEVIEW CORPORATE PARK

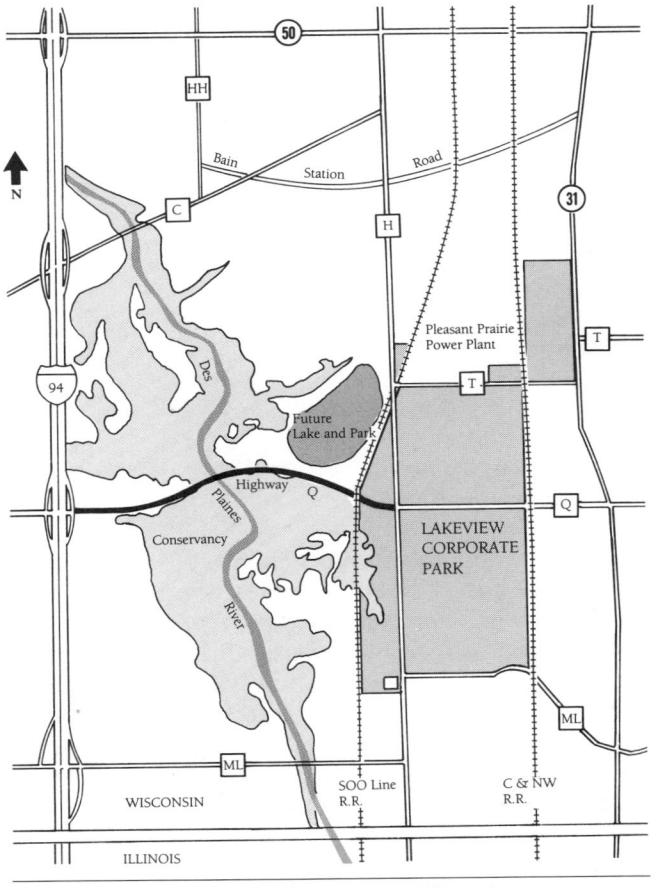

Source: WISPARK Corporation, Pleasant Prairie, Wisconsin.

FIGURE 4.6
SITE PLAN OF LAKEVIEW CORPORATE PARK

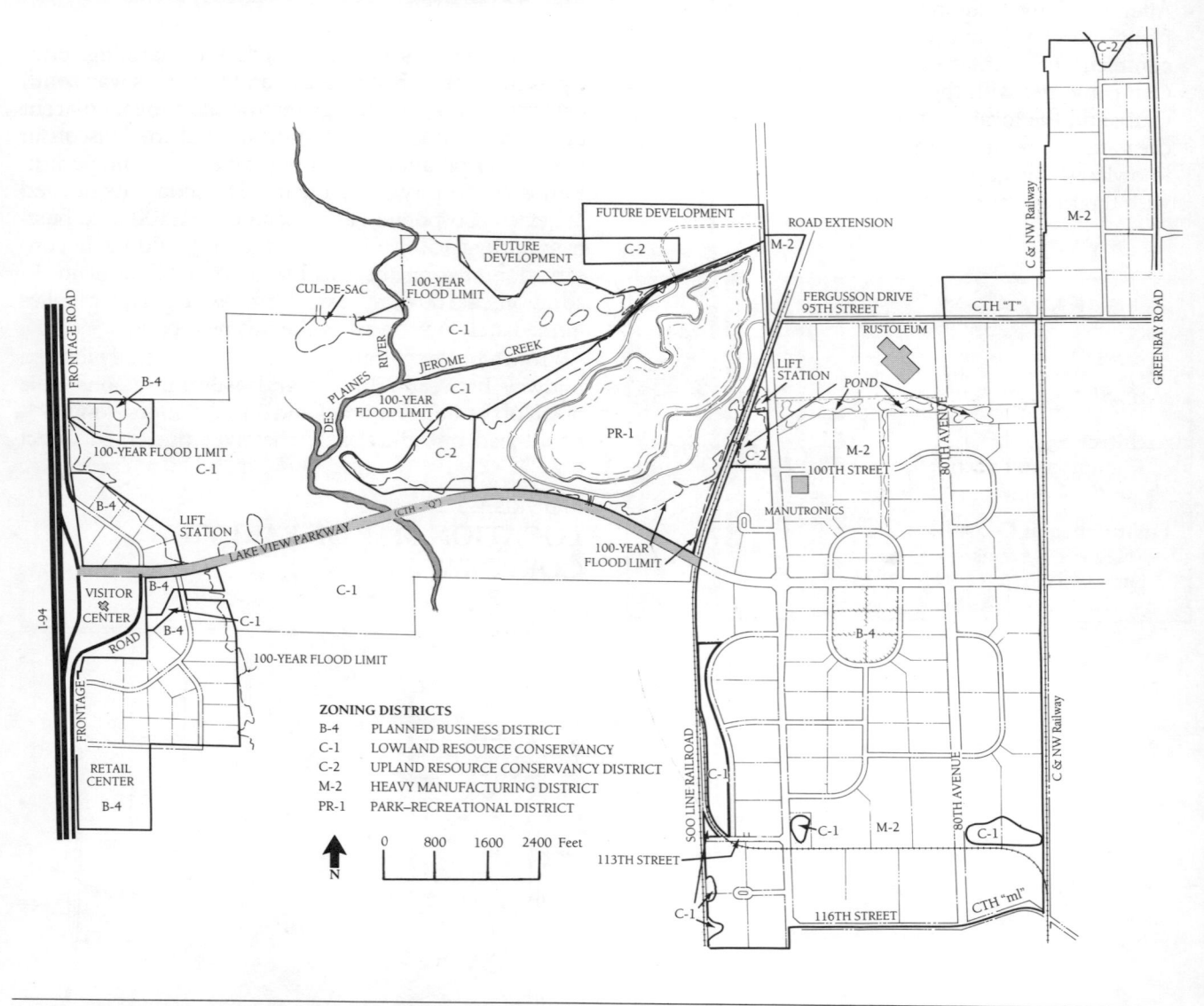

Source: Howard Needles Tammer & Bergendoff, Architects Engineers Planners, Milwaukee, Wisconsin.

access to the interstate. County Highway Q ran part way through the parcel, but stopped just short of the river. The missing link was completed in 1988 when a bridge and two miles of highway were built by the State of Wisconsin, Kenosha County, and WISPARK. The bridge and highway construction involved filling several acres of wetlands and floodplain. In exchange, the company created over 30 acres of wetlands, several acres of floodplain, and donated over 400 acres of wetlands and floodplain to The Nature Conservancy—one of the largest conservation groups in the United States. The project illustrates how conservation groups and business can collaborate to create an environmentally and economically sound development.

Minimizing Wetlands Filling

Development along the Des Plaines River corridor in northern Illinois has led to major floods downstream in both urban and suburban Chicago. Consequently, the Corps is extremely reluctant to permit any additional development in the river's floodplain or adjacent wetlands. About 10 years ago, the Wisconsin Department of Transportation proposed to extend County Highway

Q. That proposal, however, was opposed by EPA and rejected by the Corps on several grounds; in particular, there was no compelling need for the road, and the river crossing would destroy several acres of valuable wetlands and floodplain.

Learning from the Department of Transportation's mistakes, WISPARK developed a proposal in 1987 that was appealing to both EPA and the Corps. Not only would the project provide much needed jobs (a Chrysler plant had recently announced plans to close its doors), but it would also compensate for any adverse environmental impacts. In 1988, the Corps issued a permit with the following conditions: the project must not exacerbate flooding; WISPARK must establish a permanent conservation easement for its land in the floodplain, and the project must minimize wetlands impacts. WISPARK met all three conditions. It agreed to fill no more than the bare minimum (11 acres) needed to build a bridge across the river and for the corporate park itself, create wetlands and wetlands ponds out of floodplain to compensate for the fill, and establish a conservancy district for 413 acres of wetlands, wet prairie, and floodplains. Instead of the straight-line route proposed by the Department of Transportation, which would have involved filling what EPA considered high-value wetlands, WISPARK chose a route for the bridge that would circumvent most of these wetlands.

After receiving its permit from the Corps, WISPARK brought in The Nature Conservancy, a private, nonprofit, conservation organization, to carry out the restoration and manage the conservancy district. The Nature Conservancy, which was given fee title to the 413 acres, subsequently hired Applied Ecological Services, Inc. (AES), an environmental consulting firm in Juda, Wisconsin, to develop and implement the mitigation plan.

Bringing Back Fires

The existing wetlands varied from sites dominated by one or two plants such as reed canary grass to relatively diverse wetlands associated with wet prairies and floodplains along the river. Overall, the quality of the wetlands was poor.

Over the years, weeds had squeezed out native plants, primarily because natural controls, such as fires, that kept the weeds in check were absent. Historically, natural fires periodically burned the wetlands and wet prairies in the Midwest. Native plants tolerate and even depend on periodic burning, whereas exotic plants usually do not. Thus, the absence of fires allowed the weeds to proliferate, and was probably the single most important factor in the loss of species diversity (cattle grazing also played a part). In addition, the Des Plaines River was historically a pristine, gravel-bottom river. As a result of prolonged soil erosion from farms within the watershed, however, it is now a silt-laden river. Over time, thick layers of silt accumulated in the wetlands and floodplain and smothered native plant seeds, while robust weeds, such as reed canary grass, easily grew through the silt. The combined lack of fires and heavy buildup of silt gradually changed the complexion of the floodplain and wetlands along the Des Plaines River. But the mitigation plan, when fully implemented, will turn things around by bringing back some of the natural controls, like fires. Aggressive plants such as cattail and reed canary grass, which can dominate a site, are held in check when periodically burned.

In the course of this project, a total of 15.8 acres of wetlands/floodplain were filled—four for the corporate park and the remainder for the bridge. In exchange, WISPARK created out of floodplain two wetlands, one of 5.5 and one of 9.2 acres, and six small wildlife ponds totaling nine acres. The ponds vary in size from about 0.75 to 1.5 acres and are about five feet deep. They will contain a mixture of emergent, submergent, and floating vegetation and will be managed primarily to benefit flood control and wildlife. The existing 413 acres of wetlands, wet prairie, and floodplain are being restored to a system that will contain predominantly native plants. WISPARK also expanded the floodplain to compensate for the reduction in flood storage capacity that resulted from the bridge fill. For every acre-foot of flood

Once common in the prairies throughout the Midwest, fires will again be used to control exotic plants in the restored wetlands.

Buried for years by silt and a thick growth of reed canary grass, the long-dormant seeds of native prairie plants sprout anew on this wetland/wet prairie site prepared out of a floodplain. Although most revegetation will occur naturally, some seeding occurred to give the fledgling wetland a boost.

storage lost by filling wetlands and floodplain, WISPARK created an equivalent amount of flood storage by excavating an adjacent area of upland to a level equal to that of the existing floodplain.

Natural Revegetation

WISPARK's original proposal called for seeding most of the site with seeds from native wetlands and wet prairie plants. The seeds, however, are hard to come by and would cost about $600,000. Brent Haglund, then director of the Wisconsin chapter of The Nature Conservancy, now president of the Sand County Foundation, offered to do more with less. An avowed cheapskate, Haglund felt that $600,000 was too much to spend on seeds, and so the conservancy, in consultation with AES, submitted a revised proposal to the Corps.

Under this proposal, which the Corps accepted, WISPARK gave the conservancy $567,000 to manage the entire 413 acres, including the created wetlands: half of the money went into a permanent fund to pay for long-term management of the site, and the other half is being used for start-up costs associated with the restoration. Rather than buy seeds, the conservancy will sow the seeds of regeneration simply by creating the conditions that favor native wetlands and wet prairie plants over exotics. For example, in areas with thick stands of weeds such as reed canary grass, AES will scrape off the sod and expose the native, dormant wetlands and wet prairie seeds buried below. According to Steve Apfelbaum at AES, "There is a good seedbank of native plants in the substrate, so by manipulating the site, much of the revegetation will occur naturally." The Nature Conservancy will also work with upstream farmers and the Soil Conservation Service to reduce erosion and thus decrease the river's silt load. Eventually, after the native plants become fully established, the conservancy will harvest and sell the seeds produced for use at other restoration projects. The proceeds will help pay for management of the site.

Some seeding will occur in selected areas—for example, on the upland areas around the created wetlands and at one of the created wildlife ponds. In addition, the created wetlands will be planted with a cover crop of smartweed and with annual grasses such as barnyard grass to stabilize the soil until the native plants emerge and become well established. Fire will be an important management tool. Prescribed burning, which will occur about once a year for the first few years, then less so later, will keep the weeds at bay. Already, about 200 acres have been burned.

The conservancy will also conduct a modest research program on marsh and wet prairie restoration. One of the wildlife ponds created will be planted with a variety of wetlands plants, while the others will be allowed to revegetate naturally. The research program, although not on the same scale as the proposed Wetlands Research, Inc., project downstream, will monitor the natural succession of created wetlands and compare the performance of wetlands that were seeded or planted with those where Nature was allowed to run its course. In contrast to the Wetlands Research project, in which heavy construction equipment was used to reshape the environment and where the depth and flow of water will

This specially modified all-terrain-vehicle was used to plant seeds in wetlands while exerting little pressure on the spongy wetland soils.

be carefully controlled, most of the WISPARK restoration will occur naturally; there will be very little physical manipulation of the environment.

Both the Corps and EPA have stated that this project could serve as a model for development and preservation of natural areas, particularly wetlands. The only apparent problems with the project are that two of the six ponds were improperly graded; the side slopes were graded too steeply, at about 4:1, whereas the mitigation plan approved by the Corps called for a much more gradual slope, more like 8:1. The ponds will therefore be regraded. This is not unusual; most mitigation projects require a certain amount of adjustment and fine-tuning, especially during their first year or two.

Lessons Learned

- Plant cover crops to control erosion and to give desirable wetlands plants a head start over weeds.
- Prescribed burning of emergent wetlands, particularly in the Midwest, will effectively control invasion by exotic plants.
- Whenever possible, take advantage of natural revegetation.

PROJECT DATA

Developer
WISPARK Corporation
Pleasant Prairie, Wisconsin

Environmental Management
The Nature Conservancy, Wisconsin Chapter
Madison, Wisconsin

Environmental Consultant
Applied Ecological Services
Juda, Wisconsin

ENHANCEMENT

"Human interventions into nature can be creative and indeed can improve on nature, provided that they are based on ecological understanding of natural systems and of their potentialities for evolution as they are transformed into humanized landscapes."

René Dubos[4]

When a wetland that is difficult to create, such as a forested wetland, is altered, frequently the most feasible mitigation option is to create an entirely different type of wetland. Rather than try to replicate a forested wetland, which may take over 50 years, it may be prudent to create a marsh that can be established in a shorter period of time and that may be more valuable to fish or wildlife than the previous wetland. This is referred to as wetlands enhancement: selectively enhancing a wetland to boost one desirable attribute, such as waterfowl habitat, over another, such as flood control. For example, a scrub-shrub, forested wetland may be excavated and converted to a pond with a fringe of emergent wetlands plants, or a relatively common wetland could be altered to provide habitat for an endangered species.

Critics of wetlands enhancement argue that it should not be considered an equitable exchange for wetlands fills since it usually results in a net loss of wetlands. Enhancement jeopardizes less desirable (and less fashionable) wetlands traits and subjects other equally valuable wetlands qualities to the whims of the regulatory agencies or developers. For example, a state fish and game agency may support enhancing a wetland to provide additional habitat for bass or ducks, whereas a water management agency may prefer to enhance the wetland for the purposes of water supply or water treatment. In one project, a dense, forested wetland, cut in half by an access road, was partially enhanced. One half of the forested wetland was transformed into a well-landscaped detention pond, surrounded by ornamental plants, a manicured lawn, and an asphalt path. The other half was left alone. Although the natural, unenhanced wetland is more productive from a biological standpoint, residents of an adjacent development view it as an eyesore and a breeding ground for bugs, while the detention pond is seen as beautiful.

Enhancement does have its place; it can create a more diverse overall ecosystem, as illustrated by the following two projects. But like many other forms of mitigation, enhancement involves tradeoffs.

Brookshire Estates
King County, Washington

According to a former King County, Washington, planning official, just about every subdivision developer in the state of Washington wants an entrance pond that makes an entrance "statement." The developer of Brookshire Estates, a 40-acre, single-family residential development in King County, was no exception. At the entrance to the project lies about one acre of a 12-acre forested wetland. The northernmost corner of the wetland, which is dominated by western red cedar and sitka spruce, was a wet meadow that had previously functioned as a cattle pasture. When developer John Buchan first laid eyes on the site in 1985, he envisioned an entrance road featuring a grassy knoll leading down to a richly landscaped pond. The King County Planning Department, however, had other ideas. Its strict wetlands regulations, which were revised in 1989, make it very difficult to alter wetlands. The regulations state that wetlands "shall not be disturbed or altered" unless:

- the wetland does not serve any of the valuable functions of wetlands, including wildlife habitat and natural drainage; or
- the proposed development would preserve or enhance the wildlife habitat, natural drainage, and/or other valuable functions of wetlands.[5]

The county rejected Buchan's initial proposal on the grounds that it would destroy the existing, albeit disturbed, wetland.

A stand of cattails encircles most of this manmade pond; young alder, cedar, and spruce trees form a backdrop.

Buchan was able, however, to craft a compromise design that would meet the intent of the wetlands regulations while increasing the wetland's functional values. He enhanced what was considered an ordinary, relatively low-value wet prairie by replacing it with an open-water wetland that serves primarily as waterfowl habitat but also provides stormwater detention and a visual amenity. Since the existing wet prairie was the lowest spot on the property, it was the logical location for a detention pond, which is required by the county. A narrow (25-foot) berm was built between the adjacent, forested wetland, which was left intact, and the newly created open water. To tie the two wetlands together, the berm was planted with the same tree species that commonly occur in the forested wetland, such as cedar, spruce, dogwood, and alder. The berm also helps to keep stormwater runoff out of the forested wetland. Together, the berm and forested wetland form a lush backdrop to the manmade pond.

Creating the entrance pond involved excavating part of the existing wetland to create four different wetlands environments: a shallow emergent marsh (up to two feet deep), aquatic beds (two to four feet deep), shallow open water (four to six feet deep), and open water (over eight feet deep). Cattail, bulrush, pond lilies, and other species were hand-planted at appropriate elevations—for instance, cattail in the emergent zone and pond lilies in shallow open water. Since most landscapers dislike working in open water and muck, the landscape contractor developed a unique planting system which enabled workers to keep their feet dry. Plant tubers were placed in burlap bags that were weighted down with gravel and dropped off the edge of a small boat at different locations. The bags sank to the bottom of the pond where the submerged tubers later sprouted into the surrounding mud. Cattail tubers were placed in nylon mesh bags to restrain their prolific growth until other plants become firmly established.

On a geological time scale, wetlands are ephemeral environments. Sedimentation eventually causes wetlands to proceed through a number of stages: from open water to shallow marsh and eventually to upland. Wetlands enhancement at Brookshire Estates took a wet prairie a step back in time by converting it into an earlier stage of geological development. In the process, the county gained a more diverse wetland and the developer obtained the access road and entrance statement he wanted. "The wetland," remarked an official at the John Buchan Construction Company, "has tremendous marketing appeal."

The wetlands enhancement was completed in 1987, before many houses were built, at a cost of about $20,000 (not including the land, which was cheap).

Tampering with wetlands often involves tradeoffs. In this residential development, a small wet prairie became a multi-purpose pond and now provides stormwater detention, wildlife habitat, and an attractive entrance.

According to Doug Webb of Subdivision Management, Inc., which provided planning and management for the project, "You get the wetland free when you buy the good land." Thus far, the enhancement appears to be successful.

Lessons Learned

- Regulators and developers can reach a compromise more easily if they meet at the development site, where they can get a feel for the project, rather than around a conference table.
- Careful planning beforehand minimizes regulatory conflicts later.
- Acknowledge that wetlands regulations are here to stay, and learn to work with them. Either left alone or enhanced, wetlands often increase the value of a project.

PROJECT DATA

Builder/Developer
John F. Buchan Homes
Bellevue, Washington

Land Planning/Management
Subdivision Management, Inc.
Bothell, Washington

Environmental Planning
Shapiro & Associates, Inc.
Seattle, Washington

Landscaping
Tutko Landscaping and Nursery, Inc.
Redmond, Washington

Engineers
GeoDimension
Kirkland, Washington

Ostergaard-Robinson
Everett, Washington

Haig Point
Daufuskie Island, South Carolina

Haig Point is located along the South Carolina coast, on the northernmost headland of Daufuskie Island. Only five miles long and 2.7 miles wide, with three miles of ocean shoreline, the 5,200-acre island is positioned obliquely along the Savannah River entrance and Calibogue Sounds, about one mile landward from Hilton Head Island. By boat, the island is about 10 miles from Savannah. With a maximum elevation of only 30 feet, the island contains over 950 acres of wetlands.

In the early 1980s, a group of developers formed the Daufuskie Island Land Trust and bought an option on about 2,300 acres of land—which was comprised of three plantation-era tracts—from the Plantation Land Company. The trust, led by the land planning firm of Edward D. Stone, Jr., and Associates, developed a master plan for the three tracts, which are called Haig Point (1,040 acres), Oak Ridge (532 acres), and Webb (741 acres). In 1984, International Paper Realty Corporation of South Carolina (IP) purchased the Haig Point tract, with an option to buy the remaining two tracts.

Haig Point, which is being developed first, is designed to be an exclusive residential community of about 950 dwelling units on 1,040 acres. It will feature recreational amenities such as golf, tennis, and swimming for residents, members, and guests only. The market is primarily for second-home buyers. Oak Ridge is envisioned to be a mixed-use community with a golf course, a resort hotel, and a beach club, while the Webb tract will be the marine center of the three tracts. Webb will contain a marine complex that includes a 64-acre marina, a commercial area, and a major ferry terminal. The marina will feature a 94-slip river marina, a 40-acre inland harbor with 284 slips, and capacity for 450 boats in dry stack. The Webb tract will also house a golf course, tennis courts, a hotel, and a harbor-oriented village center.

Thus far, over half of Haig Point is developed, but no development has taken place on either the Oak Ridge or Webb tracts. As of January 1990, Haig Point contains 532 single-family lots and 30 multifamily units, a unique, world-class 20-hole golf course (two holes are optional), and a nine-hole golf course. The gross average density for the three tracts will be 1.26 units per acre, with a total of 2,956 units planned.

Permitting

Altogether, over 35 permits from over 20 agencies were needed for this project, including permits from the South Carolina Coastal Council (SCCC), the South Carolina Department of Health and Environmental Control, and the Beaufort County Joint Planning Commission. In 1981, the Daufuskie Island Land Trust received preliminary approval from Beaufort County for a PUD of about 2,900 units on over 2,400 acres. In 1982, the trust obtained master-plan approval from SCCC, whose jurisdiction includes salt marshes, waters of the coastal zone, critical areas, and some beachfront.

IP did not need a Corps Section 404 permit for any of its initial work at Haig Point, even though substantial dredging, particularly for the golf course construction, and some filling was involved. First of all, the Corps does not require a permit for dredging wetlands, as long as the dredged material is not deposited in wetlands. Secondly, the small amount of fill in the forested wetlands predated the Corps' expansion of its jurisdiction over "adjacent" wetlands, which occurred after the *Riverside Bayview* decision. Nonetheless, in keeping with its philosophy of environmentally sound development, IP mitigated any adverse impacts, even though there was no regulatory requirement to do so.

Built in 1872 and now restored, this lighthouse serves as a guesthouse for Haig Point.

FIGURE 4.7
LOCATION MAP OF DAUFUSKIE ISLAND

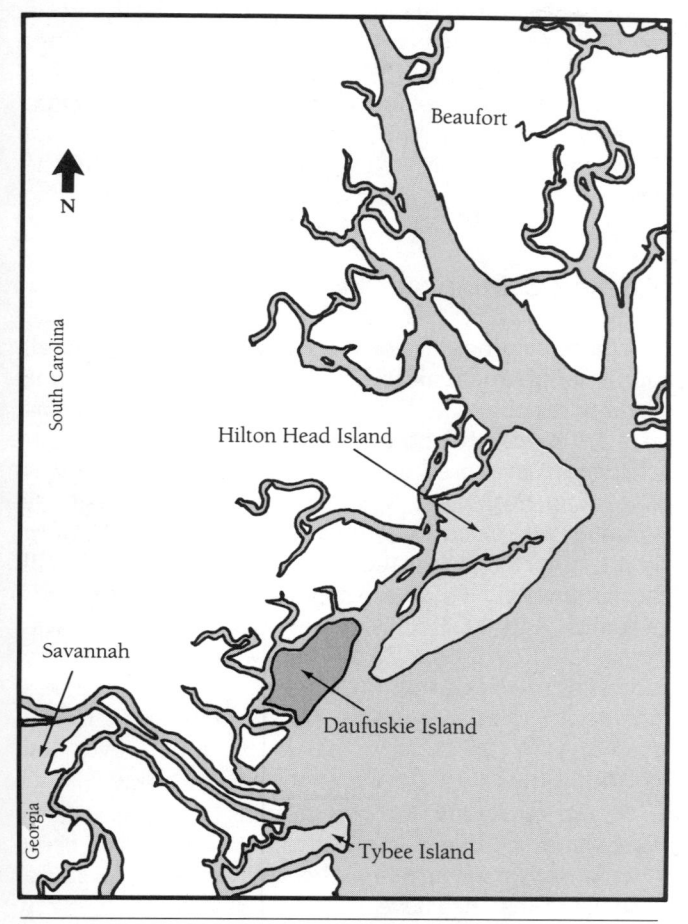

Source: International Paper Realty Corporation of South Carolina.

A Special Place

Daufuskie Island has always had a certain mystique about it. Isolated and unreachable by car, it was seen as the last island untouched and unspoiled by developers. The Daufuskie Island Land Trust knew that in order to overcome local resistance, it would have to design a project that minimized adverse environmental impacts and compensated for any damage done. IP took extra steps to preserve the historic and natural features of the island, and it shows. Elevated boardwalks carry pedestrians and golf carts from low-lying areas to the coastal uplands and protect the salt marshes below. The developer also buried all utility lines to avoid cluttering the landscape. No filling or dredging will occur in the salt marshes, and the developer compensated for all disturbances to the forested wetlands.

Probably the most important step the developer took to protect the island's fragile ecosystem was to exclude the automobile. Like Mackinac Island, Michigan, no cars are allowed at Haig Point; transportation is by horse and buggy, bicycle, electric golf cart, or on foot. In addition, IP restored a historic lighthouse, built in 1872, which now serves as a guesthouse. And a 79-year-old mansion was barged 100 miles up the Atlantic Intracoastal Waterway from its previous site on St. Simon's Island, Georgia, to serve as a guesthouse and restaurant. As a result, the SCCC praised the master plan for Haig Point as one of the most environmentally sensitive proposals it had ever reviewed. The entire island is on the National Register of Historic Places.

IP's wetlands mitigation plan fit perfectly with the construction of the golf course. A considerable amount of sand and dirt was needed to build up the dramatic tees and greens. Rather than ship it in from far away, IP excavated sand and dirt from several different locations on site and created two- to eight-acre lakes out of the resultant holes. Each lake is fringed with "littoral shelves"—shallow steps that support native wetlands plants such as torpedo grass, lizards tail, and pickerelweed. The whole project was well orchestrated. As the lakes were dug, the perimeter was graded to the proper elevation to create a littoral shelf into which native plants were hand-sprigged. Since different plants prefer different water depths, a separate zone of the shelf was created for each plant species introduced. The shelves were carefully constructed around the lakes to ensure that they would flood to a depth that is appropriate for each plant. After the lakes were dug, shelves graded, and native plants transplanted from the surrounding

Most of the holes on the golf course play over or around the wetlands. To minimize damage to the wetlands, boardwalks, like the one shown in the background, carry both pedestrians and golf carts over the wetlands.

(Before enhancement) A variety of plants such as blue flag and pickerelweed were hand planted in this freshwater marsh, which was formerly a narrow, seasonally flooded swamp forest containing black gum, sweet gum, and red maple trees.

(After enhancement) Three months later, the wetland supports a healthy stand of mixed emergents; many of the plants have already produced seeds.

marshes, the water level was allowed to rise to a predetermined level. In creating the lakes, the developer agreed to maintain a buffer of forested wetlands in order to minimize wildlife disturbances.

IP agreed to seek a comprehensive permit for all future wetlands work on the island. Under this innovative approach, all state, federal, regional, and local resource agencies could, for the first time, look at all likely cumulative aspects of IP's development proposal for its 2,320 acres on the island. During the 30-month review process, which involved numerous on-site meetings with the Corps, EPA, FWS, and several state agencies, IP reduced its request for fill from 31 acres to only eight acres. After several revisions to satisfy the concerns of the resource agencies, the Corps issued a permit for the project.

Adding Ecological Diversity

The remaining development of Haig Point and of all of the Webb and Oak Ridge tracts involves filling some eight acres of isolated, nontidal freshwater swamp forests for roads and other residential land uses. In exchange, IP proposed to create about 50 acres of open-water and freshwater wetlands out of upland areas by grading and excavating upland areas to a minimum depth of six feet below the water table. The 50 acres will be comprised of 6.3 acres of new freshwater marsh wetlands and 43.7 acres of new open-water lake systems. Unlike the island's forested wetlands, which dry up occasionally during prolonged dry spells, the lakes will retain water year-round. The new freshwater marsh system will cost approximately $140,000 over the course of the future development on Daufuskie Island.

Found primarily in long troughs between ridges, freshwater wetlands on Daufuskie Island have been altered by years of farming. The newly created lakes and associated marshes have greatly increased the diversity and numbers of animals on the island, while also increasing groundwater recharge and pushing back saltwater intrusion. IP expects that the addition of open water to the island will enhance the island's overall ecological diversity. In addition, the wetlands plants will take up any excess nutrients, including fertilizers, that are washed off the golf courses and help prevent eutrophication of the lakes—a problem that has plagued some of the lakes at nearby Hilton Head. This assumption did not, however, wash well with some of the regulatory agencies, who complained to the Corps that IP could have further reduced the amount of its wetlands conversion.

IP estimates that it spent approximately $510 per 1,000 square feet of littoral shelf created—a good investment. "The introduction of open water is good for the island environment, and even better for real estate," said Ray Pantlik, director of engineering, planning and research at IP. As of January 1990, IP had created five acres of new freshwater or salt marsh habitat at a cost of about $107,000.

Lessons Learned

- Hire an ecologist/consultant to help design the project and the mitigation. An ecologist gives the developer credibility with the regulatory agencies and suggests that the developer is making a sincere effort to protect sensitive resources. At Daufuskie Island, an ecologist was part of the initial planning team.
- Meet with the Corps before you meet with resource agencies such as the FWS and EPA. The Corps will delineate wetlands, inspect the site, and give you a good idea of the mitigation requirements.
- Do not rely on aerial photographs and large-scale wetlands maps to delineate the wetlands on your site. Aerial photographs are excellent tools for preliminary planning but cannot be relied on for final detailed planning. At Haig Point, aerial photographs indicated that the project would involve altering or filling about 22 acres of wetlands. But a field inspection revealed that only about eight acres of wetlands would be disturbed. On the other hand, a field inspection may also indicate that there are more wetlands than previously thought.

PROJECT DATA

Developer
International Paper Realty Corporation of South Carolina
Hilton Head, South Carolina

Landscape Architect and Planner
Edward D. Stone, Jr., & Associates
Fort Lauderdale, Florida

Golf Course Architect
Rees Jones, Inc.
Montclair, New Jersey

Project Ecologist
Dr. Joe A. Edmisten
Pensacola, Florida

Civil Engineers
Thomas and Hutton Engineering Company
Savannah, Georgia

Land Sciences Corporation
Fort Lauderdale, Florida

CREATION

"The whole difference between construction and creation is exactly this: that a thing constructed can only be loved after it is constructed; but a thing created is loved before it exists."

Charles Dickens, Preface to Pickwick Papers (1836)

Wetlands creation, last in the sequence of mitigation steps and last of the strategies to be discussed here, is probably first in people's minds when they think of wetlands mitigation. This strategy involves creating wetlands from scratch; turning dry woods into swamps, sandy shores into salt marshes. Creation has generated more heated debates and been the subject of more journal articles than any other form of mitigation. Why all the fuss? In general, environmental groups, resource agencies, and a number of prominent biologists oppose this form of mitigation because, they assert, it is fraught with risk and uncertainty. In contrast, wetlands creation is strongly supported by the Corps, developers, and especially by a cadre of environmental consultants who traipse around the country creating wetlands where none existed before, including mangrove forests in Florida, tidal flats in California, and freshwater marshes in Illinois—with varying degrees of success.

Building a wetland sounds easy enough. Wetlands occupy low-lying areas, contain water-loving plants, and fill at least occasionally with water. On paper, all you have to do is remove the existing vegetation, dig a hole, fill it with water, sprig in some wetlands plants, invite a few animals, and let Nature take care of the rest. In a few years, no one will ever know the difference. It is, of course, more complicated than that.

Like all natural systems, wetlands are complex, dynamic environments that have evolved over millennia. They occur in places where the natural conditions were favorable for their formation: the topography, soil conditions, hydrology, and climate were just right. The delicate interdependencies among wetlands plants and animals are not fully known and therefore difficult to recreate. Superficially, artificial wetlands may appear equal to natural wetlands, but on closer inspection subtle, but important, differences often emerge. For example, created and natural wetlands may contain similar plant mixes, but a new wetland may lack the nutrient-cycling capabilities and productivity of a natural wetland. Ideally, the created wetland should provide the same functions, such as flood control or groundwater recharge, as the destroyed wetland. Realistically, replacing all wetlands functions is difficult. Furthermore, a newly created wetland is a disturbed wetland, and weeds such as purple loosestrife and common reed thrive on disturbance. These aggressive plants can

This two-acre wetland was created by excavating three small ponds and constructing two small islands. Located alongside a highway, just outside Portland, Oregon, this manmade wetland was designed primarily for wildlife habitat.

One year after construction, the wetland was judged by the Oregon Department of Fish and Game as one of the more successful wetlands created in the state. Ironically, the state Highway Department has plans to fill much of the created wetland when it widens the adjacent highway.

quickly invade a freshwater site and crowd out more desirable plant species. In addition, even successfully created wetlands may become targets for development. In Washington County, Oregon, for example, a two-acre wetland containing three small ponds with islands was created adjacent to a highway in exchange for filling a one-third-acre wetland that had marginal value for wildlife. A few years later, the state highway department announced plans to fill the fledgling wetland to widen the highway.

One axiom of wetlands creation is that simplicity breeds success: the simpler the system, the easier it is to recreate. Marshes are easier than bogs, and salt marshes are easier than freshwater marshes. Salt marshes have proven to be one of the best candidates for replication. They present fewer variables with which to contend. Few plants have adapted to the harsh conditions of intertidal zones and few can stand the stress of water-logged roots, waves, and high salinity. Weeds present less of a problem in salt marshes than they do in freshwater marshes. Along the Delaware Bay in New Jersey, for example, two grasses dominate the salt marshes: cordgrass and salt hay. These grasses can become established in a single season. By contrast, other types of wetlands, such as forested wetlands, have a more complex hydrological regime and a greater diversity of plants and animals, take a long time to develop, and are therefore much more difficult to recreate. Forested wetlands, such as bottomland hardwood forests or cypress swamps, mature slowly and may not leave adolescence until they are 50 years old. Unlike herbaceous plants, which can grow to maturity in a single season, woody plants take longer to develop and this lengthens the time required for a wetland to become established. Noted one ecologist, "A herbaceous wetland is easy to create—a piece of cake—compared to woody marshes."[6]

Several studies have evaluated the performance of created wetlands. Many attempted to determine whether the wetlands were "successful" or not. Some based success on whether or not wetlands plants were established, some simply on whether or not a 404 permit applicant met the permit requirements, and others on how well the created wetland replaced lost functional values. For example, a 1985 study of 32 wetlands created in Virginia's coastal zone as part of a 404 permit requirement found that less than half of the replacement wetlands were successfully established. The wetlands varied in size from 300 square feet to eight acres. Success was based on a project's compliance with the permit and the condition of the created wetland. Of the 32 wetlands creation sites, only 19 were completed, four were in progress, seven had not even started, and the status of the remaining two sites could not be determined. Only nine of the 19 completed sites were considered completely successful, six sites were either partly successful or unsuccessful, and at four sites it was too early to tell. The two most common causes of failure

CREATING A SALT MARSH IN LONG ISLAND, NEW YORK

In East Patchogue, Long Island, a small coastal marsh lies along an estuarine stream that enters Patchogue (pronounced patch-hog) Bay. Originally a cordgrass marsh, the hydrology of this 2,000- by 1,200-foot wetland was severely altered when a canal, constructed prior to 1970 for a residential development, was cut through the marsh to the bay. As in many disturbed wetlands, common reed quickly moved in and covered the site.

In exchange for building a dock on the canal and a walkway through the marsh connecting a residential development to the dock, a local developer agreed to restore a portion of the former cordgrass marsh. The restoration plan, designed and implemented by Dr. Ron Abrams of Dru Associates in Glen Cove, New York, called for creating a crescent-shaped cordgrass marsh along the canal bed.

The first task was to remove the thick stand of common reed that blanketed the site and to take steps to prevent its return. Typically, herbicides are used to kill unwanted plants like common reed, but state and local regulators balked at this method, fearing contamination to Long Island's precious and vulnerable groundwater. Instead, an excavator was brought in to remove the reed grass; on its first day on the job, the heavy machine sank up to its cab in mud. It took the full force of two bulldozers to free the partially submerged excavator. Abrams then constructed a temporary roadbed to support the heavy equipment.

As well as physically removing the reed grass, Abrams dug a narrow, 12-foot-wide channel along the inner curve of the new wetlands site to keep the invasive reed grass at bay. Once the reed grass was finally removed, the area was filled with sandy loam from a nearby construction site, graded to the proper elevation, and prepared for planting.

Creating a salt marsh typically involves hand planting thousands of young cordgrass plants and scheduling the work around the tides. In this project, workers could only plant for four to five hours per tidal cycle. Thirty-one thousand cordgrass sprigs in peat plugs were hand planted at, on average, 18-inch staggered centers and sprinkled with time-release fertilizer. A crew of five people took 15 days to complete the planting. While waiting to be planted, the cordgrass sprigs, stored on site, had to be protected from predation by geese and muskrats, and also prevented from drying out in the hot July sun. The plants were soaked daily with freshwater from a nearby construction site. Only about 1 percent of the plants died in storage.

After planting was completed, a crew of two to four people closely monitored the site for three months. They checked for predation, debris that might impair drainage, and for any remaining clumps of common reed that would allow it to gain a foothold in the new marsh. Initially, geese preyed heavily on the young cordgrass, but relatively few plants died. By the end of September, the cordgrass was three feet high and covered roughly 90 percent of the site.

The total cost for the restoration, including planning, obtaining government approvals, site preparation, planting, and two years of monitoring and maintenance, was about $60,000. According to Abrams, it cost about as much to monitor and maintain the site as it did to prepare and plant it.

July: At low tide, workers hand planted over 30,000 cordgrass sprigs. Unavailable locally in such large quantities, the plants came from a nursery in Maryland.

June: The existing bed of common reed was excavated and hauled off site for disposal. To keep the excavator from sinking in the thick mud, a temporary roadbed was constructed.

September: By the end of the summer, the cordgrass had reached three feet tall in most places. Through diligent monitoring and maintenance, 99 percent of the plants survived.

AN EXPERIMENT IN VERNAL POOL CONSTRUCTION

Once common throughout the Central Valley region of California, vernal pools are now becoming scarce—a victim of agricultural development. For most of the year, these small, shallow depressions do not look at all like wetlands. Only during the rainy season do they become marshlike. Indeed, until a few years ago, the Corps did not even regulate fill in vernal pools.

Vernal pools may, however, soon become common again, in part because of their intolerance of weeds and exotic plants. In 1986, a new law required that 7 percent of federal highway landscaping funds be earmarked for planting wildflowers along highways. In California, many planting projects failed because the wildflowers were unable to compete with exotic plants. Typically, a project looked good for the first year or two but was then taken over by weeds. Vernal pools, though, are resistant to invasion by all but a few weeds because the pools present such hostile conditions of prolonged wet and dry periods. If such wetlands can successfully be created along highways, they can meet the federal wildflower requirement while providing additional habitat for endangered plants and animals, such as the delta green ground beetle.

To demonstrate the feasibility of replacing some of the lost vernal pools by creating new ones, the Department of Environmental Design at the University of California-Davis constructed eight interrelated vernal pools along a highway intersection near Davis, California. Four pools were constructed by compacting a mixture of topsoil found on site with bentonite to create a hardpan layer similar to that which underlies the topsoil of natural vernal pools. The other four pools were reconstructed from topsoil and hardpan excavated from nearby natural vernal pools on sites slated for development.

In order to study and compare which plants colonize the vernal pools and how quickly the plants migrate to other pools, two pools were fully seeded, four were partly seeded,

Most of the vernal pools constructed in this experiment were seeded not only with wildflowers but also, on their upper terraces, with native grass seed and noninvasive erosion control seed intended to restrict incursion by weeds.

and two were left unseeded. According to the designer of the vernal pools, Kerry Dawson of the Department of Environmental Design, "We wanted to see what plants colonized the pools; like throwing a party and seeing who comes."

The ponds were constructed in the fall of 1986, and by March 1987 had filled with water from the winter rains and drowned most exotic plants. But plant cover was only about 50 percent, giving weeds a chance to invade the bare ground. The fully seeded pools contained the greatest ground cover. Researchers also discovered that certain plants, such as downingia, remain where planted and do not migrate quickly to other pools. Wildflowers grew as readily in those pools where on-site soil had been mixed with bentonite as in those pools constructed from soil taken from natural vernal pools.

Four pools used hardpan from nearby natural vernal pools; the other four used a mixture of bentonite and on-site topsoil.

Although a few pools still held a small amount of water by late May of 1987, most were in full bloom with lasthenia dominating.

were inappropriate grading and an absence of erosion controls.[7]

Most wetlands scientists agree that creating wetlands where none existed before is far more difficult than restoring wetlands in their original sites. They disagree, however, about whether or not it is even possible to create a wetland, and over the criteria used to gauge success.

Wetlands creation is still more of an art than a science. The following two case studies, one from Illinois and the other from Massachusetts, illustrate some of the problems and some of the successes of creation.

NORTH–SOUTH TOLLWAY DUPAGE COUNTY, ILLINOIS

When highways and wetlands cross paths, as they often do, the result is either an immediate and total destruction of a wetland through filling, or a gradual degradation through sometimes severe changes in its hydrology. By slicing through wetlands and splitting them in two, highways can block natural water circulation and turn robust wetlands into languid pools of stagnant water. Salvaging wetlands in the path of development from imminent destruction presents an appealing, albeit problematic, solution. As this example shows, "moving" a wetland is possible, but some valuable pieces will be left behind forever.

In metropolitan Chicago, the Illinois State Toll Highway Authority is constructing a highway that is designed to relieve congestion and shorten commutes in the suburbs west of Chicago by connecting I-90 to I-55. The approximately 18-mile-long and 300-foot-wide North–South Tollway, as it is called, passes through old fields, a few residential properties, a portion of the Morton Arboretum, and over 75 acres of wetlands. Surrounded by development, most of the wetlands have deteriorated over the years. Surprisingly, however, an EIS uncovered a few wetland gems in the rough: three separate sites contained about 26 acres of extremely rare and irreplaceable wet prairie and seasonally saturated sedge

FIGURE 4.8

LOCATION MAP OF NORTH–SOUTH TOLLWAY

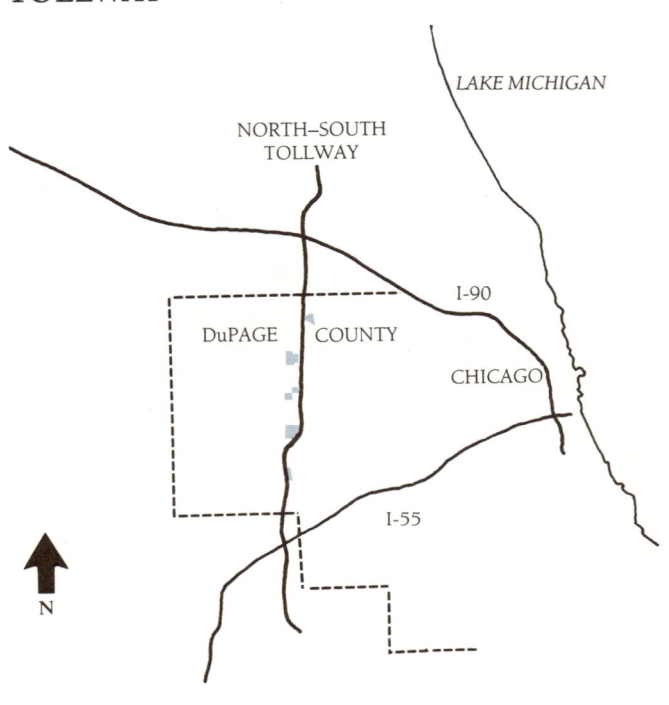

FIGURE 4.9
NORTH–SOUTH TOLLWAY WETLANDS CREATION SITES

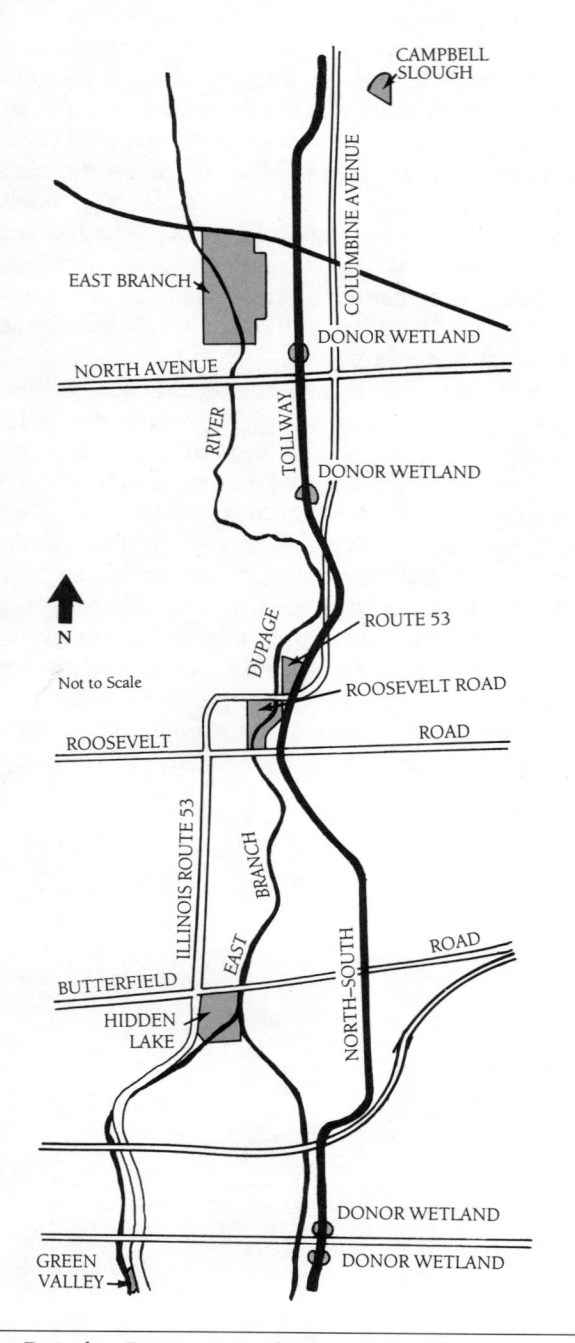

Source: Envirodyne Engineers, Inc., Chicago, May 19, 1987.

swale terrain of recently glaciated portions of Illinois. These wetlands types historically contained very high species diversity.[8]

A wetlands working group organized by the Highway Authority's consultant, Envirodyne Engineers, Inc., and composed of representatives from state and federal agencies, the Morton Arboretum, and the Highway Authority, developed a wetlands mitigation plan that would adequately compensate for the 75 acres of fill. The group developed a plan that called for creating about 120 acres of wetlands at six nearby sites, including relocating about three acres of the high-quality wetlands by moving their plants and topsoil to a new, prepared site. The plan became a condition of the Corps permit, which was issued in October 1986.

The Highway Authority's main task was to find sites that would be suitable for creating wetlands and that could also be used as a borrow pit to meet the authority's needs for over 3.5 million cubic yards of fill to construct the highway. The authority found a willing partner in the Forest Preserve District of DuPage County (hereafter, the district), which had several suitable sites nearby, two of which could be used as borrow pits and which were later converted to wetlands. Three other sites were purchased by the authority and donated to the district. The sixth site was already owned by the district. Five wetlands sites were within the floodplain of the east branch of the DuPage River, but only one site was considered a wetland, albeit a low-quality wetland dominated by reed canary grass. And all six created wetlands will be managed by the district.

Wetlands Relocation

Initially, five sites were selected for the created wetlands: Route 53, Roosevelt Road, Hidden Lake, East Branch, and Campbell Slough. A sixth site, called Green Valley, was added later. Five of the six sites are described briefly below, followed by a more detailed description of the wetlands relocation effort at the sixth, Campbell Slough.

All the wetlands creation efforts involved excavating topsoil from wetlands that were to be filled and transplanting it to the creation sites. Except in the case of Campbell Slough, the topsoil was stockpiled while the mitigation sites were prepared. After sitting for one to two years, the rich, seed-laden topsoil was spread as a top layer at each of five sites. Some members of the working group feared that many of the seeds and plants would not survive prolonged storage. After one growing season, however, enough dormant seeds in the topsoil sprouted in their new home that no planting was necessary.

meadow and shallow marsh containing over 300 species of vascular plants, including two state endangered species. Topographically, the 26 acres lie in depressions or drainageways that are typical features of the swell and

Route 53. Located directly adjacent to the North–South Tollway, Route 53 is the smallest of the five sites. When completed, the new wetland will consist of six acres of emergent wetland, 3.2 acres of open water, and about three acres of a wet prairie/sedge meadow.

Roosevelt Road. Also adjacent to the tollway, the Roosevelt Road site straddles the east branch of the DuPage River. It consists of 10 acres of open water, 10 acres of emergent plants, and about 14 acres of a wet prairie/sedge meadow.

Hidden Lake. An 82-acre site, Hidden Lake was created out of the borrow pit dug by the Highway Authority. Designed primarily for waterfowl, the site contains a 16-acre lake with about four acres of emergent fringe and 12 acres of wet prairie/sedge meadow. It also contains a small meandering creek, channelized years ago, that was restored by Envirodyne.

East Branch. A 50-acre site that also served as a borrow pit for the Highway Authority, the East Branch site was also developed primarily for wildlife. It contains 23 acres of open water, a 19-acre emergent wetland, and about four acres of wet prairie/sedge meadow.

Green Valley. A 40-acre site, with 11 acres of created wetlands, Green Valley is located about five miles south of Hidden Lake. The wetland was designed as a deep water habitat bordered with zones of emergent vegetation. Upland areas around the wetland were planted with a mixture of prairie grasses and wild flowers. Green Valley is the only site that contains a gradient from prairie to wetlands habitat.

Moving Rare Wetlands to a New Home

Relocating the rare wetlands to the five-acre Campbell Slough site involved transplanting about two-and-one-half acres of wetlands topsoil from two sites—one a seasonally wet prairie, the other a seasonally inundated sedge meadow and shallow marsh. Unlike the creation efforts at the other five sites, the topsoil taken from the two donor wetlands was *applied immediately* to Cambell Slough. Campbell Slough was previously a degraded wetland that had recently been prepared to serve as a stormwater detention basin for a 130-acre watershed that includes an adjacent residential area and a highway.

In what has been described as a salvage operation, the Highway Authority endeavored to establish a facsimile of the rare wetlands at the site by mimicking the natural water regimes typically found in the wet prairie, sedge meadow, and shallow marsh. A 10- to 30-foot-wide flat shelf surrounds the five-acre basin, which has a maximum depth of 15 feet. An outfall structure in the pond was set to create a normal pool about nine inches above the shelf elevation that, after the 12 inches of transported topsoil had settled, would keep the soil saturated to the surface for most of the year. The designers varied the shelf elevation and topsoil thickness to create a variety of hydrological conditions ranging from permanent inundation to temporary flooding. This will allow different plants to grow at different elevations.

To "move" the wetlands, about 12 inches of topsoil from two wetlands in the path of the tollway were excavated, loaded onto dumptrucks, and hauled to the Campbell Slough site about four miles away. To minimize the loss of plants, the relocation occurred in winter, while plants were dormant. Moving the wetlands, however, was not as straightforward as it sounds. For example, one of the sites was covered with ice and about two feet of standing water, so a worker in chest-waders standing knee-deep in ice-cold water had periodically to measure and confirm that the proper elevation depth was being excavated.

The newly transported, rich wetlands topsoil, laden with seeds, root stock, rhizomes, and some entire plants was unloaded and distributed over the Campbell Slough site by bulldozers and gradealls. Bulldozers pushed the

Soils from several wetlands sites were loaded onto trucks and moved to different locations nearby where new wetlands are being created.

"Walking" on wooden planks to keep it from sinking in the mud, a gradeall works its way down to the water's edge to spread the transplanted wetland soils.

After one growing season, many of the plants from the old wetland had reemerged.

frozen material to the water's edge, and final placement required the use of two gradealls; one worked at the water's edge and the second operated from mats on the submerged shelf. The first gradeall placed the material within reach of the second, which then spread the graded material across the shelf. By repeatedly relocating the mats in front of itself, the gradeall was able to "walk" its way around the entire pond shelf and work in areas that were inundated with up to two feet of water.[9] The entire operation, including excavation of the donor wetlands, hauling, and spreading, but excluding the cost of preparing the Campbell Slough site, cost about $180,000. The work was performed over a three-week period in February 1987. Management of the newly established wetland will consist of prescribed burning and manipulation of the water level.

Although no seeding or planting occurred, the entire shelf of the new site was green by midsummer. After one growing season, plants were growing on over 40 percent of the site, after two seasons, 75 percent, and after three, 90 percent. The first prescribed burn occurred in the spring of 1989.

Altogether, creation of the approximately 126 acres of wetlands cost about $8.3 million.

So far, the project appears to be doing well. Many plants from the old site have emerged from the transplanted muck. However, Gerould Wilhelm, a botanist at the Morton Arboretum in Chicago and one of the members of the work group, cautions about applying the North–South Tollway example to other projects. "It is a totally spurious notion that one can destroy a rare, native plant community and replace it elsewhere," Wilhelm observed. "That was no ordinary wetland that was destroyed, and while parts of it were saved, it can never be fully replaced."

Problems Encountered

Geese fed heartily on the newly emerging vegetation, necessitating the replanting of certain wetlands areas. In addition, beavers ate the redwood slats found in each water-control structure. At Roosevelt Road, thousands of gulls quickly claimed the budding wetland, and their acidic excrement lowered the pH value of the surrounding soil and water and killed many wetlands plants. In response, the developer planted a cover crop of barnyard grass, which grows fast and tall and is hardier than most of the other wetlands plants. Furthermore, the wetlands substrate that was moved to Hidden Lake contained rhizomes of cattails and common reed, which have become a menace to the new wetland. Finally, the Cambell Slough wetland was hit by a 500-year flood that inundated most of the wetland and damaged or destroyed many wetlands plants.

Lessons Learned

- If possible, transplant wetlands soils in winter, when plants are dormant. Avoid stockpiling; wetlands soils, which contain a variety of seeds, roots, tubers, and microbes, can become very hot and may begin to break down when stored for long periods of time.
- Transplantation of a wetland is possible, but it is also expensive.
- A wetland cannot be replicated entirely, but most of its flora can be saved.
- Regular monitoring and maintenance is an integral part of successful mitigation.
- Include water-control structures in created wetlands to maintain the proper water levels and to allow the wetland to be periodically drained for maintenance, such as prescribed burning.
- In cold climates, if the created wetland is to contain fish, fish pockets of about 10 to 15 feet deep must be dug so that the fish may overwinter without freezing.

PROJECT DATA

Developer
Illinois State Toll Highway Authority
Oak Brook, Illinois

Planning and Engineering
DuPage County Forest Preserve District
Glen Ellyn, Illinois

Envirodyne Engineers, Inc.
Chicago, Illinois

Botanist
Morton Arboretum
Lisle, Illinois

Restoration Contractor
Applied Ecological Services, Inc.
Juda, Wisconsin

WESTFORD CORPORATE CENTER
WESTFORD, MASSACHUSETTS

Located directly off Route 495 in Boston's high-tech mecca, the Westford Corporate Center consists of two 82,000-square-foot office/R&D buildings. The $13.5 million project, designed by Sasaki Associates, Inc., was a joint venture of four developers (see Project Data). Eight of the site's 18 acres contained wetlands. Sasaki, faced with the possibility of struggling through the rigorous 404 permit process, turned the wetlands from a potential hindrance into an amenity by developing a plan of closely spaced buildings surrounded on three sides by wetlands. Ten acres of the site were left as natural open space and only 1.5 acres of wetlands were filled. The filled wetlands were successfully replicated in their entirety on site. Construction started in 1985 and was completed in 1986.

About half of the site was undisturbed, including a 2.5-acre wet meadow that was an extension of an adjacent wooded swamp. The other half of the site was littered with debris, including piles of tree stumps that sank six feet deep into the wet soil and leftover gravel from construction of Route 495.

The rather undistinguished grassy area in the foreground is actually a wetland created by Sasaki Associates.

FIGURE 4.10
LOCATION MAP OF WESTFORD CORPORATE CENTER

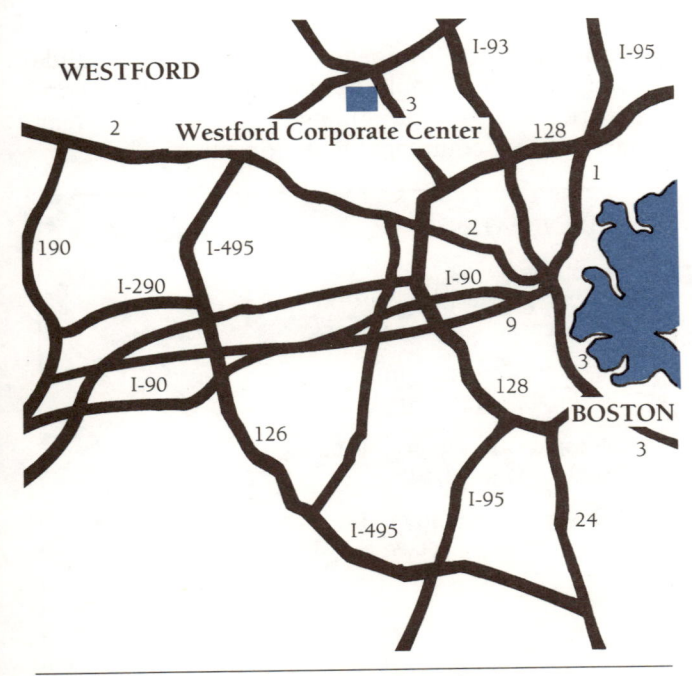

Source: Sasaki Associates, Inc., Watertown, Massachussetts.

Voluntary Mitigation

No individual 404 permit was required from the Corps since the project was covered by a nationwide permit. Under the state's Wetlands Protection Act, however, a permit (Order of Conditions) was required from the local conservation commission. Fortunately for the developers, the commission had authorized the wetlands filling for a previous owner of the site prior to the effective date of Massachusetts' regulations, which generally prohibit filling of more than 5,000 square feet of wetlands (see earlier discussion of Massachusetts's wetlands program). Under current regulations, the project's 1.5-acre fill would not be allowed. Because the fill was previously authorized, the wetlands replication was essentially voluntary, and the commission issued a permit within three months of application. No monitoring nor maintenance was required as a condition of the permit.

Although festooned with debris, the 1.5-acre wetland that was filled was a relatively simple wet meadow system and Sasaki's mitigation strategy was appropriately uncomplicated. In essence, mitigation involved moving the wetland from one place to another. The developer removed the old stumps that cluttered the replication site and excavated the area to an elevation approximately one foot below the elevation of the adjacent wetland. Excavation of the new wetland was one of the initial steps in construction of the office park. Silt fencing, installed between the edge of the adjacent

The Westford Corporate Center, shown here at dusk, stands on a site that was previously festooned with debris.

existing wetland and the replacement wetlands area, protected the existing wetland from erosion and sedimentation.

Before the 1.5-acre wetland was filled, a one-foot layer of organic material—soil, seeds, plant roots, and so on—was excavated, moved, and backfilled into the replication site, which was then graded to a slope of less than 0.5 percent from the upland edge to the edge of the existing wetland. Because proper grading is critical to the new wetland's success, engineers visited the site before the bulldozers left to ensure that the grades were correct. A mix of perennial grasses was planted to control erosion until the natural seed stock emerged from the organic muck, and silt fencing remained in place until the entire replacement wetland was stabilized with vegetative cover. Within one growing season, wetlands plants colonized the site.

The entire replication effort cost about $20,000 (less actually, because stump and debris removal would have occurred anyway). Compared to other wetlands creation projects, some of which cost over $50,000 an acre, this new wetland was a bargain.

According to Pat Loring, chairman of the Westford Conservation Commission, the new wet meadow is beautiful and "a great tradeoff for the commission." All but one of the plant species that occurred in the filled wetland, such as soft rush, tussock sedge, woolgrass, and, of course, cattail, can be found in the new wetland. Meadowsweet, however, has yet to emerge from the transplanted muck. A Sasaki engineer commented that the next time the company moves a wetland, it will stockpile certain wetlands plants and transplant them into the new area.

Lessons Learned

- Keep it simple. The more complicated the project, the more expensive and risky it becomes.
- Whenever possible, recycle the soils and vegetation in the wetland that will be filled for use in the replacement wetland. If possible, build the replacement wetland contiguous with an existing wetland.
- Make sure that the replication area is graded to accommodate the drainage needs of the wetlands plants. Grading should be done at the time of construction so that if the grading has to be corrected, all the necessary equipment is still on site.

Sasaki gave order to the two buildings by using an old European format of a tree-lined allee leading to a courtyard. With eight acres of wetlands on three sides, more than half of the site is preserved as natural open space.

PROJECT DATA

Developers
Old Stone Development Corporation
Providence, Rhode Island

The Lord Company, Inc.
Boston, Massachusetts

Gillespie & Company, Inc.
Boston, Massachusetts

Lee Kennedy Company, Inc.
Boston, Massachusetts

Architect/Engineer
Sasaki Associates, Inc.
Watertown, Massachusetts

GUIDELINES FOR SUCCESSFUL MITIGATION

Wetlands regulations have added a new dimension to the development process. Not too long ago, engineers would subdivide land without ever visiting a site—a practice that resulted in many applications for after-the-fact permits from the Corps. Now, however, the implications of a few prominent cases such as *Riverside Bayview* and *Attleboro Mall* have finally sunk in and developers will often hire a wetlands scientist early in the planning stage to determine whether or not wetlands are present, to identify the type and extent of wetlands, to design and oversee a mitigation strategy, and to help guide the proposal through the 404 permit process. Many developers stress the importance of bringing in people who have dealt with EPA and the Corps and who have a successful record with mitigation projects.

Nonetheless, wetlands mitigation is still a risky business that offers few if any rules to follow and no guarantees that a created or restored wetland will perform as planned. The absence of standardized designs means that every design must be tailored to fit the specific conditions found at each site. However, based on discussions with regulators, developers, and consultants, and after reviewing the examples presented here, some general mitigation guidelines have emerged, particularly for restoring and creating wetlands.

Avoid First, Create Last

Conformity with the sequence of mitigation steps defined by CEQ—that is, avoid all fills where feasible, minimize adverse impacts, and compensate for any damage done (see p. 32)—has proven to be both an environmentally and an economically sound approach to developing land containing wetlands. Not only is this approach advocated by most state and federal resource agencies, but it is also usually the simplest and least costly strategy. As one proceeds through the sequence from avoidance to creation, the mitigation becomes more complicated and generally more expensive.

Meet with the Corps

If wetlands are present, contact the district engineer of the Corps to determine if the site is covered by a nationwide or general permit. If an individual permit is required, applicants should also meet with other federal as well as state resource agencies, in order to understand and anticipate the regulatory agencies' positions and responsibilities. Applicants should discuss their development plans and show how they will reduce adverse impacts and compensate for any losses. The Corps is much more willing to approve a project if an applicant makes a "good faith effort" to mitigate any adverse impacts. Again, however, each district office is different. A project approved in one district may be rejected in another. Therefore, permit applicants must become familiar with how a local district operates and its position on wetlands filling and wetlands mitigation. In addition, a few developers have also suggested meeting with environmental groups to identify and address their concerns in advance and avoid the risk of project delays if opposition mounts.

Use Experienced Crews

Creating or restoring a wetland is both an art and a science. At its simplest, it involves moving truckloads of mud or sand from one place to another and covering the prepared site with wetlands seeds or seedlings. Although such a process is straightforward, the specifications must be exact; each wetland requires a specific elevation, substrate, hydrology, and plant type. Construction crews are often unfamiliar with, or unappreciative of, the complicated and delicate nature of wetlands ecology. As one wetlands consultant said, "Many contractors cannot tell a cattail from a great blue heron." One well-intentioned but misguided contractor dutifully went to a nursery to purchase the wetlands plants specified in the mitigation plan, which called for planting the site with pickleweed and cordgrass—two species that live in salt marshes. But instead of cordgrass, the nursery mistakenly gave him cattail, a freshwater plant. The contractor planted what he thought was cordgrass at the required five-foot centers and, a year later, all the cattail were dead and the site had to be replanted.

Develop a Mitigation Plan

A project design should incorporate a mitigation plan from the very beginning. Just as the architectural design of a building should "fit" the landscape, so should the mitigation design or plan. One developer consultant drafts mitigation plans with the help of a design team that includes land planners, environmental scientists, engineers, and landscape architects. In general, the more specific the plan is, the better the mitigation results will be. For example, instead of merely stating that the mitigation area will be covered with suitable wetlands plants, the plan should specify, say, that "spartina alterniflora will be hand-sprigged at six-inch intervals at the site and planting will be done in early spring." Researchers working on a study of 32 created wetlands in Virginia found that of seven projects whose permits included conditions that specified more than just the size and species of plants to use, six were successful;

> ## MITIGATION PLANNING: EXPECT THE UNEXPECTED
>
> In exchange for filling about 16 acres of forested wetlands in southern New Jersey, 32 acres of forested wetlands were created out of upland. The mitigation plan called for creating an in-kind forested wetland of primarily maple, gum, and cedar out of an upland forest dominated by red maple and black gum, including an area of cultivated blueberries and a disturbed area dominated by common reed.
>
> Creating the wetland was fairly straightforward and involved clearing the existing vegetation, scraping off and stockpiling the top three inches of soil, excavating to three inches below the seasonal high water table, replacing the topsoil, seeding with a mix of grasses and trees, and then planting over 17,500 trees and shrubs, primarily red maples. For a number of reasons, however, over half the plants were dead after one growing season and the developer had to replant over 9,000 trees and shrubs the following year.
>
> Although the mitigation plan called for planting a mix of 18- to 36-inch trees such as red maple, black gum, and Atlantic white cedar, and seeds for a number of other trees and shrubs, only maple saplings were readily available in large quantities and only from distant nurseries. Thus, most of the trees, including 15,000 maples, came from a nursery in Tennessee (seeds for pitch pine actually came all the way from Korea, via a nursery in Massachusetts). Trees from Tennessee were not well suited to growing conditions in New Jersey. In addition, the site became a popular area for off-road vehicles, which destroyed many of the young plants.
>
> The primary reason for the high mortality, however, was the combination of two stressful conditions that occurred immediately after the planting was completed: an extremely wet spring in which the young plants were exposed to standing water for a prolonged period of time, followed by a drought. Although it grows in both upland and wetlands, red maple is particularly susceptible to prolonged inundation by water. Subsequent planting substituted water-tolerant shrubs, such as buttonbush, for maples.

where the Corps established a time limit for completion, three of three were successful, but where no time limit was set, only six of twelve were successful.[10]

Every mitigation plan should contain, at a minimum, the following seven components, or, as one EPA official put it, "the seven steps to Nirvana."

1. **A clear statement of the objectives of the mitigation.** The mitigation plan should establish unambiguous goals; for example, 70 percent vegetative cover with native plants by the end of the first year, 90 percent by the second year, and 100 percent by the third. Without such a statement, the success or failure of the mitigation cannot be determined.
2. **An assessment of the wetlands values or resources that will be lost as a result of the fill and of those that will be replaced.** This assessment will provide the only way to determine whether or not the mitigation will offset the resources or values lost from the wetlands alteration.
3. **A statement of the location, elevation, and hydrology of the new site.** Hydrology is the most important factor in creating a wetland. If grades are just slightly off plumb, plants will either be left high and dry or flooded. Water must be provided and controlled either by excavating down to the water table, by collecting and routing stormwater runoff to a wetland, or, as in the Ballona wetlands restoration, through a system of pumps, weirs, and gates. In one poorly designed project, a consultant installed plants inappropriate to the site's hydrology. Instead of regrading and replanting the site, the consultant simply installed a sprinkler system just to keep the plants alive—expedient yes, but not exactly successful mitigation.
4. **A description of what will be planted where and when.** Mitigation failures commonly arise from selecting the wrong plants for a site, selecting poor-quality plants, or planting at the wrong time. Studies have shown that local, indigenous plants do better than plants grown and shipped in from a distant greenhouse and that sprigs do better than seeds.
5. **A monitoring and maintenance plan.** Monitor completed projects to ensure that the objectives are met and maintain the site to keep the wetland functioning properly. Specify who will monitor, how often, and who will report on and review the results.
6. **A contingency plan.** Almost every created or restored wetland will require periodic adjustments, especially during its first year. A number of things can and often do go wrong. For example, one mitigation project involved transplanting over 23,000 wetlands plants as part of an effort to create four intertidal estuarine islands. Much of the planting occurred at night, when the tide was low. A subsequent site inspection revealed that about 30 percent of the plants had been planted upside down, and a crew had to be sent out to turn them over. Other, more typical problems are:

Weeds. Weeds can be a problem at any site. Just as dandelions will invade the most manicured lawns, cattails, purple loosestrife, and common reed will invade restored or created freshwater wetlands. And they can quickly take over a site. They can be controlled, however, by establishing a crop of quick-growing annual grasses to cover the site and give the wetlands plants a head start on weeds, by periodic burning to kill the weeds but not native plants, or by limited applications of herbicides.

Planting a cover crop of annual grasses can backfire, however, as one disappointed consultant discovered. The consultant planted rye grass to stabilize the site and reduce invasion by weeds, but the rye proved more competitive than expected and still covered most of the site 15 months after it was planted. Confusingly, although most states require that plants such as cattail and reed canary grass—considered serious pests in some states—be controlled at restored or created wetlands, other states actually encourage these and other weeds to be planted because they provide such a quick, dense plant cover and prevent erosion.

Disturbance. In Washington, a state official has found that wetlands created adjacent to industrial areas fare much better than those created near residential developments. The reason? Children like to play in wetlands—digging holes, building dams, leaving trash, riding motorcycles—and their activities can wreak havoc with a wetland. For example, in one mitigation project in California, birds that came to feed and nest in a newly created marsh were harassed by people playing with remote-controlled toy airplanes. Wetlands near industrial areas face less human disturbance because they are usually located far enough away from where people live and play.

Predation. Muskrats, geese, and other animals can destroy a newly planted marsh. In one project, a consultant carefully planted 150 young trees in a forested wetland that was being restored. Soon after, beavers ate every tree. Some consultants install owl "scarecrows" to frighten away small birds and animals that eat young plants.

Availability of plants. Few nurseries stock wetlands plants such as cordgrass or mangrove seedlings. In addition, some states now limit the practice of "borrowing" wetlands plants from one site to use at another. The unavailability of plants may delay implementation of a mitigation plan.

A contingency plan should specify what procedures will be followed should things go awry.

7. **A guarantee that the work will be performed as planned and approved.** Some states, such as California, require that the mitigation be implemented before or concurrent with construction of a project. Once the buildings on a site are constructed, developers may run out of interest, time, and, most importantly, money to complete the mitigation. Bonding, which has long been used to guarantee delivery of promised infrastructure improvements, is now being used to ensure that mitigation work will be carried out as promised and that any flaws in a wetland will be corrected. Florida regulators can require developers to post a performance bond for mitigation projects that will cost over $25,000 or if the developer has a poor track record.

Notes

1. Personal conversation on June 2, 1989, with Mike Knowlton, deputy zoning administrator of Fairfax County, Virginia.
2. San Francisco Bay Conservation and Development Commission, "Mitigation: An Analysis of Tideland Restoration Projects in San Francisco Bay," an unpublished staff report, dated March 1988, p. 3.
3. John J. Zentner, "Wetland Restoration Success in Coastal California," in *Wetlands: Increasing Our Wetland Resources. Proceedings of a National Wildlife Federation Conference Held on October 4–7, 1987*, eds. John Zelazny and J. Scott Feierabend (Washington, D.C.: National Wildlife Federation, April 1988), p. 216.
4. René Dubos, "Symbiosis Between the Earth and Humankind," *Science*, vol. 193, August 1976, p. 460.
5. See 21.54.170 King County Code.
6. Personal conversation on October 18, 1988, with Dr. Joe Edmisten, an ecologist living in Pensacola, Florida.
7. Christopher O. Mason and Dean A. Slocum, "Wetland Replacement—Does it Work?," in *Coastal Zone 87: Proceedings of the 5th Symposium on Coastal and Ocean Management* (New York, New York: American Society of Civil Engineers, 1987), p. 1183. (The study was done by Maguire, C.E., Wetland Replacement Evaluation Contract DACWG5-85-D-0068, U.S. Army Corps of Engineers, Norfolk District.)
8. Sue Elston, John Rogner, Gerould Wilhelm, and Wayne Lampa, "Establishment of a Native Midwestern Wetland Plant Community Through Relocation of Marsh Topsoil," in *Proceedings of a National Workshop on the Beneficial Uses of Dredged Material* (Vicksburg, Mississippi: Environmental Laboratory, U.S. Army Engineer Waterways Experiment Station, November 1988), pp. 150–155.
9. See Elston, et al., "Establishment of a Native Midwestern Wetland Plant Community," p. 153.
10. Mason and Slocum, "Wetland Replacement—Does it Work?," p. 1183.

CONCLUSION

The difficulty lies, not in the new ideas, but in escaping from the old ones.
John Maynard Keynes[1]

Throughout history, people have altered wetlands to create harbors, control mosquitoes, expand farmland, and provide homesites. In the conterminous United States alone, about 100 million acres of wetlands—about half of what existed when the first European settlers arrived—have been destroyed. The United States still loses between 300,000 and 500,000 acres of wetlands each year from both manmade and natural causes. A wave of new legislation has, however, slowed the rate of wetlands losses. Incentives that once encouraged draining and filling wetlands have been replaced by restrictions. Some of the most powerful incentives rewarded farmers who converted wetlands to cropland. Today, farmers think twice before draining wetlands, thanks in large part to the swampbuster provisions of the 1985 Food Security Act. And developers, while not necessarily jumping on the "save-our-wetlands" bandwagon, have significantly reduced wetlands losses by avoiding, restoring, or creating wetlands.

Yet, despite developers' best efforts to steer clear of wetlands, more filling is bound to occur as communities seek room to grow. How can communities accommodate growth without destroying wetlands? A growing number have adopted wetlands protection ordinances that set tough standards for wetlands development. For example, in Hillsborough County, Florida, developers must provide acre-for-acre replacement for a filled wetland, periodically monitor and maintain the artificial wetland for several years, and designate the site as a permanent conservation area. King County, Washington, regulates activities in wetlands as well as on steep slopes, along streams, and in floodplains under its "sensitive areas ordinance." Under the ordinance, dredging or filling wetlands is not allowed without a permit from the county, which inventoried its wetlands and classified them according to size and value. In general, the county designated low-value wetlands under one acre as class 3, wetlands over one acre as class 2, and wetlands over five acres as class 1. Development is permitted in class 3 wetlands as long as no net loss in wetlands acreage occurs, but development is essentially prohibited in class 1 and 2 wetlands. In addition, the county also requires buffers of 100, 50, and 25 feet for class 1, 2, or 3 wetlands respectively.

Several communities rely on special area management plans to shunt development away from high-quality wetlands and to provide developers with a more predictable permit process.

States have also been active in protecting wetlands. Over the last 10 to 15 years, several states have adopted their own wetlands protection laws. Typically broader and stricter than their federal counterparts, state wetlands laws cover more than just dredging and filling and include, for example, draining and excavation. Maryland's 1989 Nontidal Wetlands Protection Act represents one of the most recent examples. The act regulates removal, excavation, and dredging of any material from wetlands, as well as filling or draining wetlands. More states, particularly in the humid East, will likely follow suit as they strive to control their wetlands losses.

Several states have developed wetlands acquisition programs. For instance, Iowa, which has lost more than

90 percent of its original wetlands, spends over $2 million a year to acquire, restore, and manage wetlands, primarily those that were drained for agriculture. The state now restores between 700 and 800 acres of wetlands per year. In addition, a growing number of private groups, such as Ducks Unlimited, are buying and restoring wetlands.

At the national level, the trend of expanding federal jurisdiction over wetlands is likely to continue. And with their newfound authority to issue administrative penalties, both EPA and the Corps will be more likely to pursue and fine wetlands scofflaws. Moreover, in contrast to their antagonistic past, EPA and the Corps have recently shown a greater willingness to cooperate. Their landmark agreement, drafted in late 1989, on a mitigation policy is the latest in a number of mutual agreements on regulating wetlands. The agreement formally acknowledges the agencies' mutual goal of following the sequence of mitigation steps—avoid first, minimize second, and compensate as a last resort—when reviewing permit requests, and embodies the concept of no net loss of wetlands, a concept previously endorsed by President Bush.

Implementing such a policy, however, has proved troublesome. Twice the White House delayed the implementation date of the agreement in response to objections raised by real estate and oil interests, particularly those in Alaska. A watered-down version of the agreement was finally released on February 7, 1990. The revised agreement maintained the agencies' no-net-loss goal but added a provision that allows those who propose to destroy wetlands to skip the established sequence of mitigation steps if an area contains a "high proportion of land which is wetlands."

Even if the agreement between the Corps and EPA had been approved in its original form, however, it would not have reduced wetlands losses appreciably. Likewise, even if the Corps and EPA were to prevent all illegal fills, they would reduce the total amount of wetland losses only slightly. By most estimates, the Corps regulates only about 20 percent of activities that destroy wetlands. The remaining 80 percent—draining, dredging, and ditching—occur legally. And although several states have enacted stricter regulations to pick up some of the slack, state laws often contain many of the same agricultural and forestry exemptions as the federal program. If the nation were serious about protecting wetlands, it would pressure state and federal legislators to enact legislation that plugs such loopholes.

Congress might also consider stiffening both the Corps' and EPA's enforcement capabilities. The agencies simply do not have the resources, nor some say the will, to enforce the regulations adequately. In the San Francisco Bay Area, for instance, the Corps has not resolved two-thirds of the 60 cease and desist orders it issued against property owners who illegally filled wetlands between 1985 and 1990. And while a few prominent wetlands enforcement cases in late 1989 and early 1990 caught the attention of the development community, enforcement still remains lax. Probably the only thing more annoying to a developer than spending considerable time and money complying with the 404 regulations, is to see competing developers ignoring the laws with impunity.

It might be argued that the courts have proven better protectors of wetlands than have EPA and the Corps. Although local, state, and federal laws regulating activities in wetlands have generated controversy and more than their share of lawsuits, they have withstood nearly all legal challenges. Typically, property owners assert that the laws constitute a taking of property without compensation, a violation of the Fifth Amendment of the U.S. Constitution. Courts have generally struck down such challenges on the basis that either the mere existence of a law or the designation of property as wetlands is not a taking, or that plaintiffs must be denied *all* economic use of their property before a taking can occur.

The consequence, whether intended or not, of many wetlands laws is to make development in wetlands such a hassle that developers will take their bat and ball and go play elsewhere. Several court cases, particularly the *Attleboro Mall* decision, underscored the uncertainty of the development process. As a result of that decision, developers must pass a very imprecise test before qualifying for a 404 permit. They must demonstrate that no practicable alternatives to wetlands filling existed at the time they entered the market, whenever that might be. The possibility of a prolonged, uncertain permit process and extensive mitigation requirements have deterred many developers from building in wetlands.

Although in an ideal world, development would never occur in wetlands, in reality, some wetlands filling will occur and new wetlands will be built or old ones refurbished to compensate for the loss. Wetlands preservation and development are not necessarily mutually exclusive, as some of the examples in this book have illustrated. Developers have shown that they can mitigate adverse impacts of their projects—not always and not in all wetlands, but certainly in some. Examples abound of where developers avoided most, if not all, wetlands alterations by carefully choosing the location of their buildings, roads, and utilities and by controlling erosion and stormwater runoff into wetlands during construction. Increasingly, developers have demonstrated that they can improve the environment by restoring degraded wetlands, including some of the most polluted and abused wetlands, such as those in the Hackensack Meadowlands. And although not all of the wetland's original values can possibly be replaced, most people would welcome such restorations.

One crucial unanswered question about mitigation is whether or not wetlands can be successfully recreated. So far, the evidence suggests that the answer will be both "Yes" and "No." Certain wetlands, such as salt marshes and mangrove swamps, are easier to create than others. Such manmade marshes come closest to replacing all the functions and values of natural wetlands. For most other wetlands types, such as bogs and bottomland hardwood swamps, however, the jury is still out. Usually, some of the more obvious functions, like wildlife habitat or flood control, of just about any wetland can be replaced. For instance, waterfowl are notoriously indiscriminate about where they rest and feed, and are as likely to be seen on a water-filled gravel pit pond as on a pristine natural marsh. But what of the less obvious values, such as nutrient recycling, groundwater recharge, or plant productivity, that defy easy detection or measurement? Should artificial wetlands be judged by appearance only? If wetlands plants, birds, and other animals are present in about the same relative mix and numbers in a new wetland as in the natural one, is that a successful project?

One wetlands biologist aptly cited urban renewal to illustrate the criteria that must be applied in evaluating the success of artificial wetlands: "When you restore an ailing city center, it's not successful once the buildings are renovated. People must move in; business must attract consumers; profits must be made; crime rates must fall. It's not successful until functions are restored. And it won't be persistent unless some compatible mixture of residential, commercial and industrial activities is provided."[2] Perhaps EPA and the Corps should develop standards for development in wetlands to reflect the relative difficulty of recreating different wetlands types—strict regulations for bogs, more lenient for salt marshes. If we treat all wetlands equally, we are likely to end up with a very limited range of wetlands types.

While federal, state, and local agencies and governments work to establish mitigation standards, ongoing wetlands research projects may shed new light on different mitigation techniques and help eliminate some of the current uncertainties associated with creating new wetlands. For example, the Wetlands Research, Inc., project along the Des Plaines River will help unravel some of the damp, dark secrets of wetlands ecosystems. In addition, the Corps has been conducting its own wetlands research for years through its waterways experiment station in Vicksburg, Mississippi. The experiment station designed the restoration of Point Mouillee Marsh as well as several other wetlands. EPA has recently initiated a number of studies to review the performance of manmade wetlands nationwide. And following the National Wetlands Policy Forum's recommendation to expand substantially federal research on wetlands creation and restoration, both the Corps and EPA will probably increase their research efforts.

The results of these and other projects, however, will not be forthcoming for several years. In the meantime, development marches on and new discoveries about wetlands restoration and creation will continue to arise from developer-sponsored mitigation projects. Such projects have taught important lessons about constructing wetlands, such as establishing the proper hydrology, using native plants, controlling weeds, and protecting fledgling wetlands from disturbances. One successful and now widespread wetlands creation technique is to salvage as much of the existing wetland as possible and move it to its new location: in practice, this entails scooping up an eight- to 10-inch layer of muck containing seeds, rhizomes, and roots of wetlands plants, as well as nutrients, organic matter, and invertebrates, hauling this mixture to the site of the new wetland, and spreading it around the properly graded new site. This technique was successfully applied at several projects, including the Westford Corporate Center and the North–South Tollway.

In some parts of the country, off-site mitigation sites have become a scarce commodity. In the San Francisco Bay Area, for example, it is almost impossible to find an available, affordable site for off-site wetlands mitigation. Most of the shoreline is either very expensive, has already been developed, or contains a perfectly good wetland that does not need restoring. This puts regulators in a difficult position of either denying permits for projects that are otherwise in the public interest, approving them without mitigation, or allowing the mitigation to be located far off-site and with perhaps a different type of wetland.

Mitigation banking offers one possible solution to this dilemma. Despite some initial setbacks, mitigation banks will probably become more popular, both because of their convenience and also because they provide a reasonable solution to the problem of mitigating small, isolated wetlands fills. Several states that have explicitly authorized mitigation banks, but have not yet used them, may activate them in light of the National Wetlands Policy Forum's endorsement of mitigation banks.

For better or worse, wetlands mitigation is here to stay. Like most new fields of study, much of what is learned is by trial and error. Some techniques have proven effective and have been used to create or restore wetlands throughout the United States. Others involve significant risks and should be used with restraint until the problems can be worked out. As yet, no standard criteria to determine wetlands values and no consistent mitigation guidelines have emerged—although EPA is working on establishing such guidelines. With no standards and few models to guide would-be mitigators, that developers and their consultants have been successful at all is equally commendable and surprising.

What lies ahead? Developers should anticipate facing tougher mitigation standards as well as greater demands for compensation for projects in wetlands. Regulators will be expected to become familiar with the different approaches to development in wetlands and able to identify those mitigation techniques that have proven effective and those that have not. And defenders of wetlands can take some comfort from the fact that an increasing number of developers are restoring tired and tattered wetlands, leaving healthy ones intact, and, only after all other options have been exhausted, carefully moving those that are in the way.

For the moment, those who are considering developing in wetlands should bear three things in mind:

Location counts. Many state and local governments have enacted their own wetland laws, which vary in stringency. Activities allowed in one state or county may be forbidden in another. And coastal wetlands tend to be more vigorously regulated than inland wetlands.

Abide by the rules. To the uninitiated, the federal 404 permit process may seem like a labyrinth of bureaucratic requirements. Indeed, the permit process can be tortuous and complex for those insensitive to the Corps' and, in particular, EPA's interests in preserving wetlands. Generally, however, the process is fairly straightforward. Before applying for a permit to fill wetlands, developers should conduct a thorough search of alternative, non-wetland sites and make every effort to mitigate adverse environmental impacts of a proposed project. If federal or state agencies object to what they consider avoidable wetlands filling, developers may find themselves suffering through extensive permit delays and expensive redesigns.

Approvals far exceed denials. As a whole, the nation still favors development over preservation of wetlands—judged, at least, by the laws and programs designed to regulate development in wetlands. The Corps still approves a substantial majority of the applications it receives to fill wetlands. Granted, many of these permits require some sort of wetlands creation or restoration in exchange for fill, but the fact remains that the Corps seldom denies a permit to fill wetlands. And EPA rarely uses its 404(c) authority to veto a Corps permit.

Notes

1. From Keynes's preface to *The General Theory of Employment, Interest and Money* (London: 1936), p. viii.
2. Joy B. Zedler, "Why It's So Difficult to Replace Lost Wetland Functions," in *Wetlands: Increasing Our Wetland Resources. Proceedings of a National Wildlife Federation Conference held on October 4–7, 1987*, eds. John Zelazny and J. Scott Feierabend (Washington, D.C.: National Wildlife Federation, April 1988), p. 121.

BIBLIOGRAPHY

The American Law Institute. *Wetlands: Development, Progress, and Environmental Protection under the Changing Law.* Philadelphia: author, 1987.

Baldwin, Malcom. "Wetlands: Fortifying Federal and Regional Cooperation." *Enivironment*, vol. 29, no. 7, October 1987.

California Coastal Commission. "Statewide Interpretive Guidelines. December 16, 1981." (Available from California Coastal Commission office in San Francisco.)

The Conservation Foundation. *Protecting America's Wetlands: An Action Agenda.* The Final Report of The National Wetlands Policy Forum. Washington, D.C.: author, 1988.

Cowles, Demming C., et al. "State Wetland Protection Programs: Status and Recommendations." Report prepared for the U.S. Environmental Protection Agency, Washington, D.C., December 1986.

Elston, Sue, John Rogner, Gerould Wilhelm, and Wayne Lampa. "Establishment of a Native Midwestern Wetland Plant Community Through Relocation of Marsh Topsoil." In *Proceedings of a National Workshop on the Beneficial Uses of Dredged Material, October 27–30, 1987.* Vicksburg, Mississippi: Environmental Laboratory, U.S. Army Engineers Waterways Experiment Station, Vicksburg, Mississippi, November 1988.

Feierabend, J. Scott, and John M. Zelazny. *Status Report on Our Nation's Wetlands.* Washington, D.C.: The National Wildlife Federation, October 1987.

Ferren, Wayne R., Sr., and David A. Pritchett. "Enhancement, Restoration, and Creation of Vernal Pools." Environmental Report no. 13, December 16, 1988, Environmental Research Team, University of California, Santa Barbara. Printed by UCSB Printing and Reprographic Services.

Florida Department of Environmental Regulation. "Report to the Legislature on Permitted Wetlands Projects. October 1, 1987–September 30, 1988." February 1989.

General Accounting Office. *Wetlands: The Corps of Engineers' Administration of the Section 404 Program.* Washington, D.C.: author, July 1988. GAO/RCED-88-110.

Goldman-Carter, Jan, Thomas Kean, et al. "Point Counterpoint." *Environmental Forum* (a journal of the Environmental Law Institute), vol. 6, no. 1, January/February 1989.

Hammer, Donald, and Robert Bastian. "Wetland Ecosystems—Natural Water Purifiers?" Unpublished report of the U.S. Environmental Protection Agency, Washington, D.C., and the Tennessee Valley Authority, Knoxville, Tennessee, 1988.

Harlow, William M., Ellwood S. Harrar, and Fred M. White. *Textbook of Dendrology. Covering the Important Trees of the United States and Canada.* 6th edition. New York, New York: McGraw-Hill Book Company, 1979.

Hook, D.D., et al., eds. *The Ecology and Management of Wetlands: Vols. 1* and 2. Portland: Timber Press, 1988.

Kentual, Mary E., et al. "Trends and Patterns in Section 404 Permitting in the Pacific Northwest." Report for the U.S. Environmental Protection Agency, Corvalis Research Laboratory. Forthcoming in *Environmental Management.*

Kosowatz, John. "Wetlands Establish Their Worth." *Engineering News Record*, October 1987.

Krohne, James, Jr. "When It Comes to Wetlands, There's Nothing Like the Real Thing." *Planning*, February 1989, pp. 4–9.

Kusler, Jon, Dr. *Our National Wetland Heritage: A Protection Guidebook*. 4th edition. Washington, D.C.: Environmental Law Institute, 1987.

Kusler, Jon A., et al., eds. *Proceedings: National Wetland Symposium: Mitigation of Impacts and Losses*. New Orleans: Association of State Wetland Managers, Inc., May 1988.

Landin, M.C. *Engineering and Design: Beneficial Uses of Dredged Material*. Washington, D.C.: Department of the Army Corps of Engineers, Offices of the Chief of Engineers, 1986.

——— and Andrew C. Miller. "Beneficial Uses of Dredged Material: A Strategic Dimension of Water Resource Management." In *Transactions of the 53rd North American Wildlife and Natural Resources Conference, 1988*. Louisville, Kentucky: The Wildlife Management Institute, Washington, D.C., 1988.

Larson, Joseph S. and Christopher Neil, eds. *Mitigating Freshwater Wetland Alterations in the Glaciated Northeastern United States: An Assessment of the Science Base*. Amherst, Massachusetts: The University of Massachusetts, September 1986.

Maltby, Edward. *Waterlogged Wealth: Why Waste the World's Wet Places?* London and Washington, D.C.: International Institute for Environment and Development, 1986.

Mandelker, Dan. *Land Use Law*. 2nd edition. Charlottesville, Virginia: The Michie Company, 1988.

Mason, Christopher O., and Dean A. Slocum. "Wetland Replacement—Does It Work?" In *Coastal Zone 87. Proceedings of the 5th Symposium on Coastal and Ocean Management*. New York, New York: American Society of Civil Engineers, 1987.

Oregon Division of State Lands. "Astoria Airport Mitigation Bank, Astoria, Oregon. Resource Credit Evaluation" Unpublished report, November 1986.

———. "State Assumption of the Federal 404 Permit Process." Staff Report, December 1988.

Richter, Douglass B. "Genesis and Cleansing in the Wetlands." *The New American Land Magazine*, September/October, 1987, pp. 27–30.

Riddle, Elizabeth. "Mitigation Banks: Unmitigated Disaster or Sound Investment?" California Coastal Conservancy Staff Report, October 1986.

San Francisco Bay Conservation and Development Commission. "Mitigation: An Analysis of Tideland Restoration Projects in San Francisco Bay." Staff Report, March 1988.

———. "1988 Annual Report." Report issued by BCDC, San Francisco, January 1, 1989.

———. "Diked Historic Baylands of San Francisco Bay." Staff report, 1982.

Short, Cathleen. "Mitigation Banking." Biological Report 88(41), U.S. Fish and Wildlife Service, Research and Development, Washington, D.C., July 1988.

Strickland, Richard, ed. *Wetland Functions Rehabilitation and Creation in the Pacific Northwest: The State of Our Understanding*. Olympia, Washington: Thalassaco Science Communications, 1986.

Teal, John and Mildred. *Life and Death of the Salt Marsh*. New York, New York: Ballantine Books, Inc., 1969.

Tiner, Ralph W., Jr. *Mid-Atlantic Wetlands: A Disappearing Natural Treasure*. National Wetlands Inventory Project. Newton Corner, Massachusetts: Cooperative Publication of the U.S. Environmental Protection Agency and the U.S. Fish and Wildlife Service, June 1987.

Tiner, Ralph W., Jr., U.S. Fish and Wildlife Service. *Wetlands of the United States: Current Status and Recent Trends*. National Wetlands Inventory Project. Washington, D.C.: U.S. Department of the Interior, Fish and Wildlife Service, March 1984.

Travis, William. "A Comparison of California's Coastal Programs." In *Coastal Zone 87. Proceedings of the 5th Symposium on Coastal and Ocean Management*. New York, New York: American Society of Civil Engineers, 1987.

Tripp, James, T.B., and Daniel J. Dudek. "The Swampbuster Provisions of the Food Security Act of 1985: Stronger Wetland Conservation if Property Implemented and Enforced." *Environmental Law Reporter*, May 1986, pp. 10120–10122.

U.S. Army Corps of Engineers. "Point Mouille Final Supplemental Environmental Impact Statement." Report prepared by Corps' Detroit District, Detroit, Michigan, September 1984.

U.S. Army Corps of Engineers, U.S. Environmental Protection Agency, U.S. Fish and Wildlife Service, and U.S. Department of Agriculture. "Federal Manual for Identifying and Delineating Jurisdictional Wetlands." Washington, D.C.: U.S. Government Printing Office; an Interagency Cooperative Publication, January 10, 1989.

U.S. Department of the Interior, U.S. Fish and Wildlife Service, National Ecology Research Center. "Mitigation Banking." Biological Report 88(41), July 1988.

U.S. Department of Transportation, Federal Highway Administration. *A Method for Wetland Functional Assessment, Vol II*. Springfield, Virginia: National Information Service, March 1983.

U.S. Environmental Protection Agency. *America's Wetlands: Our Vital Link Between Land and Water*. Washington, D.C.: author, February 1988.

———. *Report on Use of Wetlands for Municipal Wastewater Treatment and Disposal*. Washington, D.C.: author, October 1987.

———. *Wetland Mitigation Effectiveness*. Wakefield, Massachusetts: Metcalf & Eddy, February 1986. Report prepared for U.S. EPA by Metcalf & Eddy.

U.S. General Accounting Office. Wetlands: *The Corps of Engineers' Administration of the 404 Program*. Washington, D.C.: author, July 1988.

U.S. Office of Technology Assessment. *Wetlands: Their Use and Regulation*. Washington, D.C.: United States Congress, Office of Technology Assessment, 1984.

Valentine, Lieutenant-Colonel Kit J. "Corps of Engineers: Mitigation Responsibilities." *In Mitigation of Impacts and Losses. Proceedings of the National Wetlands Symposium*. Berne, New York: Association of State Wetland Managers, May 1988.

Zedler, Paul H. "California's Vernal Pools." *National Wetlands Newsletter*, vol. 11, no. 3, May–June 1989, pp. 2–5.

———. "The Ecology of Southern California Vernal Pools: A Community Profile." Report prepared for U.S. Fish and Wildlife Service, Biological Report 85(7.11), May 1987.

Zedler, Joy B. "Why It's So Difficult to Replace Lost Wetland Functions." In Zelazny and Feierabend, eds., 1988.

Zelazny, John, and J. Scott Feierabend, eds. *Wetlands: Increasing Our Wetland Resources. Proceedings of a National Wildlife Federation Conference held on October 4–7, 1987*. Washington, D.C.: National Wildlife Federation, April 1988.

Zentner, John. "Wetland Projects of the California State Coastal Conservancy: An Assessment." *Coastal Management*, vol. 16, 1988, pp. 47–67.

———. "Wetland Restoration Success in Coastal California." In Zelazny and Feierabend, eds., 1988.

APPENDIX

COMMON AND SCIENTIFIC NAMES OF PLANTS MENTIONED IN THE TEXT

Common Name	Scientific Name	Common Name	Scientific Name
alkali bulrush	*Scirpus robustus*	Pacific cordgrass	*Spartina foliosa*
arrow-arum	*Peltandra virginica*	pepperbush	*Clethra alnifolia*
arrowhead	*Sagittaria latifolia*	pickerelweed	*Pontederia cordata*
bald cypress	*Taxodium distichum*	pickleweed	*Salicornia virginica*
barnyard grass	*Echinochloa crus-golli*	pitch pine	*Pinus rigida*
black gum	*Nyssa sylvatica*	pitcher plant	*Sarracenia purpurea*
black mangrove	*Avicennia nitida*	pond cypress	*Taxodium nutans*
blackrush	*Juncus jerardii*	popcorn flower	*Plagiobothrys undulatus*
blue flag	*Iris versicolor*	purple loosestrife	*Lythrum salicaria*
bog laurel	*Kalmia polifolia*	red mangrove	*Rhizophora mangle*
Brazilian Pepper	*Schinus terebinthifolius*	red maple	*Acer rubrum*
cattail	*Typha* spp.	reed canary grass	*Phalaris arundinacea*
common reed	*Phragmites communis*	salt hay	*Spartina patens*
cordgrass	(see also salt marsh cordgrass)	saltgrass	*Distichlis spicata*
		salt marsh cordgrass	*Spartina alterniflora*
cranberry	*Vaccinium macrocarpon*	smartweed	*Polygonum* spp.
downingia	*Downingia ornatissima, pulchella, bella*	soft rush	*Juncus effusus*
		sweet bay	*Magnolia virginiana*
fox sedge	*Carex vulpinoidea*	sweetgum	*Liquidambar styraciflua*
glasswort	*Salicornia* spp.	torpedo grass	*Panicum repens*
Humboldt Bay owl's clover	*Orthocarpus castillehoides* var. *humboldtensis*	tupelo gum	*Nyssa aquatica*
		tussock sedge	*Carex stricta*
lizards tail	*Saururus cernuus*	Venus-flytrap	*Dionaea mucipula*
mangrove	See black mangrove and red mangrove	water hemp	*Acnida cannabina*
		wax myrtle	*Myrica cerifera*
meadow-foxtail	*Alopecurus howellii*	willow	*Salix* spp.
meadowsweet	*Spiraea latifolia*	woolgrass	*Scirpus cyperinus*
northern white cedar	*Thuja occidentalis*		